At the Distant Meridian

Memoirs of a participant in the Spanish Civil War

Modern Wars Series

Nikolai Gerasimovich Kuznetsov

Cover artwork: Republican Navy battleship Jaime I off Cartagena during the war.

At the Distant Meridian

Memoirs of a participant in the Spanish Civil War[1]

N.G. Kuznetsov

Translated by Celestino Corraliza

---

[1] The literal translation of the original subtitle text is "Memoirs of a participant in the national revolutionary war in Spain", the above is utilized to avoid Soviet era expressions which read clumsy to contemporary readers.

Published by Trident Publishing
Miami, Fl USA

Library of Congress Cataloging-in-Publication Data

Kuzentsov, Nikolay Gerasimovich
Originally published Na Dalekom Meridiane
Moscow 1968
At the Distant Meridian
ISBN-13: 978-0-9963157-7-7
Spain -- History -- Civil War, 1936-1939 -- Participation, Russian.
Spain -- History -- Civil War, 1936-1939 -- Personal narratives, Russian

Printed in the United States of America

Translator's Forward

Nikolai Gerasimovich Kuznetsov, (1904-1974), created an extraordinary career in eventful times. Joining the navy during the Russian Civil War in 1919 he rose to his final and highest rank in 1955, as Admiral of the Fleet of the Soviet Union.[2] He accumulated considerable combat, administrative and political experience. He authored a number of works relating to his life and profession. For the purposes of this work, Kuznetsov is the most important non-Spanish naval voice from the Spanish Civil War. Hitherto this work his experiences, other than his native Russian, have only been available in the Spanish work "Bajo la bandera de la España republicana: recuerdan los voluntarios soviéticos participantes en la guerra nacional-revolucionaria en España"[3] where the salient points of his chapter are often quoted in secondary works respecting Republican Spain naval operations.

The current work includes considerably more detail as is expected of a book length treatment vs. chapter summary in the aforementioned work. Kuznetsov wrote regarding Spain in the early-mid 1950s while his memoirs were fresh.[4] In his memoir we see the shaping of the Spanish navy as it grapples with the loss of most of the officer corps at the onset of revolution. Kuznetsov writes for his Russian audience and provides details regarding Soviet advisors in other military professions-notably aviators and tankers. He is also writing in an era of Communist influence and there is no doubt from his comments he is a loyal communist which colors his analysis of others.
N.G. Kuznetsov served as senior naval advisor to the Republican Navy during his tenure in Spain. His professional judgments are

---

[2] Equivalent to five star rank in the US Navy.
[3] Itself a translation of the original Russian "Pod znamenem Ispanskoĭ Respubliki, 1936-1939."
[4] Na Dalekom Meridiane was published in 1968.

vivid in detail and provide an extraordinary view into the situation behind Republican lines relating to both purely military decisions and political interests in naval affairs. Naval aspects of the Spanish Civil War have had no coverage in English language works. This highly regarded memoir fills in a needed area.

The memoir is a straightforward translation of his Russian text without deletions.[5] Kuznetsov's style is matter of fact with an occasional humor or introspective aspect. He is rather modest of his own achievements which were considerable and earned him considerable standing among the Spaniards he came into contact with.[6] The illustrations are those included in the original work.

Explanatory notes are provided where relevant to clarify a point which was unknown at the time the text was written or for phrasing now too ancient for the contemporary reader.

Every effort has been made to faithfully represent the original text and transliterations of Russian and Spanish names into English. Naturally any errors in the translation are of my own making although without intent.

C.Corraliza

Do I need to remember this?

---

[5] Other than the subtitle noted there were remarkably few, to the Western ear, stilted Soviet type phrases.

(Instead of the foreword)

The 25th anniversary of the outbreak of war in Spain was drawing near. Reminding this to my comrades - former volunteer sailors, I invited each of the participants in these events to recall and write about those memorable days. The argument caught fire. "Is it worth remembering that it is already more than obscured by other, larger events," one of those present said, and explained that it was more important to describe and draw experience from the events of the Great Patriotic War. "But after all, the civil war on the Iberian Peninsula immediately preceded the war in Europe, and then the World War," his neighbor objected. A few more remarks, and the topic was exhausted - a more urgent and closer in time overshadowed these "pages of the past." However, recalling the past and, above all, the years of World War II, I involuntarily transferred mentally to Spain. After all, there republican Spain, together with our volunteers, tried to stop the onset of fascism. It was there that the real danger of an imminent big war arose.

The intervention in Spain was the first step on the road to war, and the Spanish people were the first victims of the fascist offensive in Europe. It depended on the outcome of the struggle in Spain whether Hitler would unleash a new aggression. Less than half a year separates the end of the tragedy in Catalonia and the defeat of the Spanish Republic from World War II.

That is why I have always linked my thoughts about the Second World War with the Spanish Civil War.

Therefore, I decided to write memories of the struggle against fascism in Spain, where I was first as a naval attaché, and then, during the war, became the chief naval adviser.

30 years have passed. The intervention of international fascism helped Franco establish a dictatorship in Spain. Then, in 1936,

---

[6] Kuznetsov was nicknamed "He who commands" among the Spanish.

there was still no powerful socialist camp and the peoples of the world could not come out against the forces of war and react as a united front. As for England and France, the governments of these countries, instead of solving the political problems that arose radically, preferred to pacify Hitler and reassured themselves with the thought that the aggression of German fascism would be directed to the East. Hitler's "Drang nach Osten", it seemed to them, would allow him to avoid a war in Western Europe and get off with small concessions. Chamberlain willingly went onto this bait in Munich in 1938. We observed the same thing during the negotiations of the military missions of the USSR, England and France in August 1939. As is now known, only a treaty between England and France with the USSR against advancing fascism could serve as a real obstacle. But it was precisely this step that our negotiating partners did not want to take. They looked for an alternative where it was not, and found different plans to push Hitler to the East. In this regard, there were serious internal disagreements in both England and France, which weakened the position of these countries in the fight against fascism.

History very soon, and in the most ruthless way, paid these short-sighted politicians. Less than two weeks after the 1939 talks, England was forced to enter the war with Hitler. Chamberlain's prediction of "50 years of peace" did not come true. The French government, which calmly looked at fascism, was forced to sign surrender in June 1940.

It took more than 5 years of war and 50 million victims to defeat fascism and save civilization. But there was an opportunity to prevent the tragedy.

Years passed, much has changed, but so far there are enough manifestations of revenge and military aggression in the world.

Thinking about the events in Spain in 1936-1939, you come to the conclusion that it's too early to make them history.

This was what I was guided by when I was preparing the second edition of the book “At the DistantMeridian”. The book is supplemented with new materials, including materials about our volunteers - pilots and tankers, which will allow the reader to more fully understand the picture of the struggle of republican Spain.

They say that history repeats itself. “No, the story of fascism should never be repeated,” we must declare to those who dream of fascism.

Table of Contents

At the Distant Meridian

To the aid of the Spanish Republic
Unexpected Appointment

A rebellion against the republican government in Spain broke out on July 18, 1936. This was the beginning of the civil war - the almost three-year struggle of the Spanish Republic against internal reaction and open intervention by fascist Germany and Italy. This struggle has attracted the attention of the whole world. The progressive people of many countries voluntarily went to the Iberian Peninsula in order to fight against the fascist threat with weapons in their hands.

Our volunteers - pilots, tankers, artillerymen and sailors - also took an active part on the side of the democratic forces of Spain.

After the war, many memoirs and studies appeared, but all this literature deals with land combat, it describes in detail the main operations on all fronts involving aviation, tanks and artillery, and only a few foreign authors talk about military operations at sea. Meanwhile, the peninsular position of Spain made both warring parties dependent on military operations at sea. In the current political situation, when France, bordering with Spain, refused to sell and deliver the necessary weapons to Spain, there remained the only way to get help from the Soviet Union and other countries was by sea.

I happened to be a witness and participant in the Spanish Civil War during the year. Having been appointed to the post of naval attaché, I arrived in Madrid at the end of August 1936. The situation forced me to abandon established etiquette and generally accepted norms of diplomatic behavior. There was a struggle, and active help was required. The USSR did not remain neutral; it openly declared its readiness and even the duty to "render all possible assistance" to the fighting Spanish people. That was what determined my behavior. When I left for Cartagena a few days after arriving in Madrid, it was already clear to me that the "special conditions" that I was told about in Moscow consisted in a more active participation in the actions of the republican fleet than that for an attaché in normal peace conditions. However, ignorance of the language, perforce, initially limited my actions.

In mid-September, I took part in the campaign of the fleet to the North, which helped me get on track, and after returning, I took up the duties of chief naval adviser. I had, together with the command of the republican fleet, to organize operations to meet transports and escort them to Cartagena. The combat effectiveness of the fronts depended on the successful delivery and unloading of arriving aircraft, tanks and guns. All aviation and tank units operating at that time on the fronts received weapons through Cartagena.

During military operations, with the permission of my superiors, I went to sea as a "volunteer", on the cruiser "Libertad" or on some destroyer.

After returning from Spain, I decided to write about this period, but first, the turbulent situation in the Far East, and then just current affairs prevented me. Later, using my notes, I returned to this work, setting myself the modest goal of telling about the actions of the Spanish sailors and our Soviet volunteer sailors during the years of the national liberation struggle of the Spanish people against fascism.

Until 1936, we knew little about Spain. Cognition was limited to significant historical events and more related to the distant past. All sailors knew about the "Invincible Armada", which, consisting of 130 ships, went to sea to stop the growing influence of England. This was the last (and, it must be said, unsuccessful) attempt to maintain its former dominance on the waterways. The "Invincible Armada" since then reminds one of the power of the Spanish fleet in the 16th century and at the same time about the beginning of the decline of a rich feudal country during the period of bourgeois revolutions.

In our schools the Battle of Trafalgar was studied and, as a warning to posterity, the Franco-Spanish fleet led by Villeneuve reproached, not without reason, for poor preparation and indecision; it was this that allowed the English fleet, under the leadership of Nelson, to win.

We knew from the history of the Russian fleet that our squadrons more than once called at Spanish ports, heading from the Baltic to the Mediterranean Sea, and lingered there for a long time, using this time for combat training.

In recent years, having no diplomatic relations with Spain, we received little information about the internal situation in this country. Therefore, it is not surprising that the first reports of the rebellion launched on July 18, 1936 in Spain remained almost unnoticed. But not even a week had passed before the attention of all the Soviet people was riveted to these events. Already in early August in Moscow and in other cities, demonstrations were held in support of Republican Spain, fundraising was carried out to help Spanish fighters and the civilian population. It became apparent that events on the Iberian Peninsula were unfolding beyond the borders of one country; German and Italian intervention threatened global peace. The struggle in Spain took a protracted nature. Soon we learned that the entire republican fleet - then I said "all" - remained on the side of the government and actively operates in the Gibraltar area.

We sailors were interested in the situation at sea. In the evening, in our free time, we gathered to discuss events in Spain. On the table was a map of the Iberian Peninsula, where the front line was marked in red pencil. At first glance, the position of the rebels seemed hopeless: only the south and certain areas in the north-west of the country were captured by the rebels, and the rest of the territory was controlled by the government. It was difficult to get a clear picture of the situation at sea, so the conversation was mainly about the composition of the Spanish fleet. According to the English reference "Jane's", in the Spanish fleet there were 2 battleships, 7 cruisers and more than 20 destroyers. Spain had submarines. Individual ships, including the newest cruisers "Canarias" and "Baleares", were still unfinished in construction. Which ships remained at the disposal of the rebels was not clear. One only believed that since the rebels failed to make a coup in the capital and the largest cities, such as Barcelona and Bilbao, the victory of the government was obviously ensured.

On August 19, 1936, signalman V.V. Billevich brought me a telegram from the fleet commander I.K. Kozhanov, which in itself was already unusual. I quickly ran through the text and understood nothing: "You are permitted to leave for Moscow today. Fleet Commander". I exchanged glances with the signalman. It seemed to me that these were "communication jests," that is, some kind of mistake crept in, and he apparently believed that I knew why I was allowed to travel to the capital. In fact, I knew no more than he did. Therefore, I again and more carefully read the short telegram. It was not clear why the commander orders me, the commander of the ship, bypassing my superiors? Why do I urgently need to leave for Moscow today? Why is it "permitted" and not ordered to go, moreover, without specifying the reasons for such a hasty business trip. However, it was necessary to act. I could still catch the evening train, but needed to get confirmation from my immediate superior, without whose knowledge I could not leave the ship and go to Moscow. If the brigade commander were right there in the roadstead, I would take a boat, go to the authorities and resolve all doubts. But he was somewhere at sea, and I sent an urgent radio, in the meantime, preparing for the transition to Sevastopol. I could assume anything but the true purpose of the challenge. The time for travel in August is the most inappropriate. Until the fall, I had no intention of going anywhere: I liked commanding the cruiser "Chervona Ukraine"; having sailed on cruisers for about nine years and commanding the ship for three cruises, I coped with my duties without much difficulty. Every trip was already a definite pleasure. There were no extraordinary incidents over the past months on the ship, which means that there was no need to talk with me about this to the high authorities in Moscow. Therefore, the last thing I thought about was that having temporarily surrendered the cruiser, I would no longer return to it. Without waiting for an answer, I ordered up anchor and head for Sevastopol. As if on a well-worn road hundreds of times, the ship moved confidently, without slowing down, approaching the bay.

I did not recognize anything specific in Sevastopol. Judging by the fact that the fleet headquarters had booked tickets for the evening train for me, it was clear that the fleet command had certain indications about my urgent departure to Moscow. The response received from the brigade commander confirmed my departure, but did not bring clarity. Then I took the initiative, turning directly to the commander for instructions. "Come, I will be at headquarters," the Fleet Commander answered me and hung up. I decided to first transfer the command of the ship to my senior assistant I. D. Eliseev and prepare a small suitcase so that I could go ashore at the right time.

The deputy commander for political affairs A. Zubkov wished me a happy journey and a speedy return. Orderly Shevchenko collected the most necessary things, polished buttons on a new, "less shiny" tunic (the old tunic shone from long wear) and even changed the white cover on the cap. I then lived as a bachelor and there was no great need to call into the apartment. Everything needed for the trip was in the cabin.

The boat took me to Grafskaya Marina.

Intrigued by such an unusual call to Moscow, I was in a hurry to see the fleet commander, who, I thought, knew what and why they were calling me. The well-known building of the headquarters of the fleet stood on a hill, from where the bays and approaches to Sevastopol were perfectly visible. Not without excitement I went up to the second floor. The fleet commander received me without delay. From the cards lying on the table one could guess that the command was preparing for training. I.K. Kozhanov got up and silently walked around the office, apparently still busy with his thoughts. He was an outstanding person, and one cannot help but say a few words about him, slightly deviating from the topic.

Ivan Kuzmich Kozhanov enjoyed well-deserved authority. The February Revolution found him as Midshipman, and in March 1917 he had already joined the ranks of the Communist Party. During the Civil War, Kozhanov was the head of the detachment of sailors of the Volga Military Flotilla on the Eastern Front. It was he who once told on ship of how, while still very young, he fought on land and, having ordered "attack", was surprised that no one was rising, only then he realized that he was the first to get up and attack the enemy."Well, then everything fell into place," he then finished the story, the continuation of which we already knew. He fought bravely all the years of the civil war, and of the detachments of Ivan Kozhanov, as familiarly, but with pride they called him then, we, very young commanders, knew. In 1921, Kozhanov already commands the Baltic Fleet. Then he studies at the Naval Academy, actively participates in the discussion of naval problems. After the Academy, he was appointed a naval agent (attaché) to Japan, although he really did not want this. Indeed, it was not in his character to be away from his homeland and navy. At the first opportunity in 1930, Kozhanov asked to be appointed senior assistant officer to the senior assistant to the commander[7] of the destroyer "Uritsky", so that, not chasing after his official position, from the basics he would go through the necessary experience on ships. Soon he becomes the commander of the destroyer and behind the samovar (which happened on the destroyers) conducts conversations far beyond the scope of this post. This is not difficult to understand if we recall that in those years a fleet was created. Thoughts about which ships should be built in the first place in order to successfully fight a stronger enemy at sea, occupied the leadership. We (the Black Sea Fleet) got to know I.K. Kozhanov more closely when he was appointed commander of the Black Sea

---

7 Senior Assistant to the Commander is an officer rank from the inter-war Soviet Navy. In this case, Kozhanov is the executive officer corresponding to a Lieutenant in other navies.

Fleet in 1931. As the commander of the cruiser "Chervona Ukraine", where the commander constantly kept his flag, I often observed his work and heard his comments during the exercises. Not being an eloquent speaker, he could point out errors and explain his point of view with utmost clarity in operational games or in the analysis of exercises. Inquisitiveness and a great desire to bring something new and the most advanced in the methods of dealing with a probable adversary were the main feature of his character. This was my last meeting with him. I.K. Kozhanov was unreasonably accused and arrested.[8]

But back to the trip to Spain. "Why do you think you're being called to Moscow?" The commander asked me at a time when I myself was ready to ask about it. He frankly admitted that he did not know and did not even know the reasons for my call, but he advised me to be ready to clearly report to the authorities about the progress of the combat training of the ship. Contrary to expectations, a very short conversation took place, and I had no choice but to ask permission to go to carry out the received order. In vain, then, deep down I suspected the commander of insincerity, he did not want to initiate me into this secret. Now I know for sure that he really did not know anything about the reason for my call. I was convinced of this when I arrived in Moscow and, without wasting time, went to the head of the Naval Forces V. M. Orlov. The Naval Forces Directorate, then still part of the People's Commissariat of Defense, was located at the new building in Antipyevsky Lane. Management occupied several floors and "coexisted" with the aircorps, located, as was supposed by the nature of the work, higher than sailors. I have known Vladimir Mitrofanovich Orlov for quite some time. The difference in official position has always been great. We were two years in the same party organization on the ship, and more than once I had occasion to conduct informal, almost comradely conversations with him. While still a cadet of the naval school in 1924, I first met him as the foreman of the political circle, which was once a week led by a member of the party since 1918 V. M. Orlov. He was then the head of the naval educational

---

[8] Executed 1938. A victim of the Stalinist purges of the office corps.

institutions. It was well remembered because Orlov often talked to us after classes on various topics. He loved to know the opinion of the interlocutor, to convince him and to answer questions with his usual eloquence. As a student, he was repeatedly arrested for propaganda, and once in the midshipman school, he continued to be interested in politics and campaigned against tsarism. Not without reason on the cruiser "Bogatyr", when the revolution started, Orlov was immediately elected to the ship committee. Soon he was appointed editor of the newspaper Red Fleet, and then head of the political department. In his file he was described: "He has great organizational skills, treats his duties conscientiously, without, however, feeling a special inclination to serve in the naval department." As a curiosity in the same description it is written: "Military happiness is indefinite." By the time of the meeting described, V. M. Orlov had already come a long way in the Soviet Navy. He commanded the Black Sea Fleet for several years, and for more than five years he served as chief of the Naval Forces. I cautiously and with some uneasiness expected a meeting with high authorities. In addition, I knew that Orlov was not averse to letting go of some sharpness in your address or asking a tricky question. The study turned out to be large, and approaching the table, it was possible to get accustomed to and evaluate the situation. "Here," I thought, "finally everything will become clear to me. All dots above the "i" will be placed. And quickly. " But again, instead of talking specifically about the purpose of my call, the authorities were interested in my service on the ship, asking general questions. I was embarrassed when the conversation ended with an order to expect an early call to K. Ye. Voroshilov. If the Fleet Commander Kozhanov directly and frankly said that he did not know what was the matter, then the Chief of the Naval Forces V.M. Orlov, apparently, considered it inconvenient to confirm his ignorance and only diplomatically smiled at my request to clarify the riddle. And the casket just opened[9]: V. M. Orlov also did not know why the commander of one of the ships was urgently needed. I

---

[9] An expression from the fable of I. A. Krylov "Casket". Meaning a problem which appeared difficult had a simple solution.

was convinced of this when a few hours later I again got into his office and reported on what had happened.

I didn't get to see People's Commisar of Defense K. Y. Voroshilov, but was accepted by his adjutant R. P. Khmelnitsky and sent to one of the departments.

"Do you know what events are happening now in Spain? - the head of the department asked me a question. "Would you like to go there?" He continued, and, without waiting for an answer, he began to paint me a picture of the events taking place there, clearly exaggerating. He asked me to think before giving an answer, and emphasized that this entirely depends on my desire. I especially did not need to think. Not knowing yet my role and the tasks with which I was invited to leave for Spain, I agreed.

"Then sit down and talk in more detail. You are appointed there as our naval attaché," the head of the department began, and remarked that my compliance is required by the special conditions in the country, which is to leave urgently.

Then he stood up and carefully examined me, said: "I want to ask you what you know about Spain?" I was confused by his question, and he, apparently noticing this, added: "Well, of course, like a sailor, and no more ", - and, apologizing, withdrew "for several minutes", leaving me alone. In front of me lay a map of the Iberian Peninsula. Only the largest events surfaced in memory. Several Columbus expeditions in search of a mysterious and rich India, of course, have not been forgotten since school. Then I tried, from memory, to follow the further development of Spain in the 16th century. The "invincible armada", as the crown of the once powerful fleet of Spain, was well known. However, it seemed to me that all these facts are so well-known that they will not satisfy my examiner. "Apparently, he is interested in events related to Russia and the Soviet Union," I thought, and began to recall later times. On the map, I noticed a small port of Vigo, with which some of the Russian fleet's voyages from the Baltic to the Mediterranean were connected. It was known from naval history how our squadrons of sailing ships stood in this port for months, they replenished fresh supplies there and even intensely engaged in combat training in anticipation of decisive battles with the Turks. Our ships also often visited Cadiz. There in 1817 there was even a transfer of five large 74-gun ships and three frigates sold to the Spanish government. "This is an important fact," I thought, as a student awaiting a call to the board. However, events of the beginning of the 20th century were more clearly depicted. Well, what about Spain in the last days that preceded the rebellion? I knew that King Alfonso XIII was overthrown; that at the beginning of 1936 the government of the Popular Front came to power and, perhaps, nothing more. At this time, the head of the department returned and, while not turning to me for an answer, sat down at the table and opened one of the drawers. "Is it not for this that he left me alone to check my quickness," I thought, and recalled the anecdotal story of one of the Academy's lecturers. According to him, some "cunning" boss invited a candidate for work abroad to his office and left him alone with a request to write the necessary statement, and when he returned, the first question was: "What is in the middle drawer of the table?" "Well, so what do you remember?" Interrupted my interlocutor of these seditious thoughts. I began to tell in chronological order, like a student, stretching time. In a word, I knew little about Spain and found myself in the position of the student who was asked about Spain in the exam, and he, not knowing what to say, decided to start his story with Portugal. Instead of Spain, I had the facts about Italy that we knew better in those years. But I needed to talk about the events on the Iberian Peninsula, about the Spanish fleet. "What do you know about the composition of the Spanish fleet?" The next question was raised. I breathed a sigh of relief. This was well known in the directory, which we looked at more than once after the start of the rebellion. It's still not very clear to me what point was awarded to me for my answer, but after that the head of the department started talking to me as a naval attaché. One way or another, everything turned out smoothly, and now, having listened, I had to go to Paris at the first opportunity, and then look for a way to get to Madrid.

These were the first days of the actual restoration of diplomatic relations between the Soviet Union and Republican Spain.[10] A few days ago, our first ambassador, M. I. Rosenberg, left for Paris, followed by military attache V. E. Gorev.

"The conditions in which you have to work are unusual," my interlocutor continued, unfolding a land map of the Iberian Peninsula. He briefly introduced me to the situation on the fronts. Almost nothing new compared to what I already knew. Touching on the situation at sea, he pointed to Gibraltar, where, according to his data, the republican fleet operates, and to the ports of Malaga and Cartagena, where he is based.

The head of the department advised me not to stay too long in Madrid and as soon as possible go to where the republican squadron will be located. He painted a rather rosy picture of the suppression of the rebellion and expressed hope for an early and complete victory of the republican government. Details were proposed to be learned from our ambassador and military attaché.

Upon learning that I should buy at least some civilian dress, the head of the department postponed the departure time for a day.

As I was ordered, I went to the head of the Naval Forces V. M. Orlov. Now I already knew why I was summoned to Moscow, but he still remained in the dark.

"Report," politely, did not order, but rather asked V. M. Orlov. He dialed the phone, and I told how I came to K. Y. Voroshilov, how Khmelnitsky met me, and how then I went to the department I needed.

"Don't drag it," apparently Orlov wanted to say, but he was silent, and I continued my report.

---

[10] Author's note in the original. Formally, they were established in 1931 after the overthrow of the monarchy, but the exchange of ambassadors did not take place.

"To whom did you hand over the ship?" Vladimir Mitrofanovich asked when I finished my message with the fact that I had only one day to equip and say goodbye to friends.

When everything became clear, Orlov leaned back in his chair. In front of me was again that slightly important chief, as he had become in the last years of his service.

We politely said goodbye.

In Paris!

The Soviet plane of the Russian-German company Deruluft taxied to the airport building in Moscow and turned off the motors. We announced our route number and offered to prepare for landing. A more thorough than usual verification of documents on a foreign line, and we headed for the queue to the plane, handing over our small suitcases. Of the nine passengers, eight were Soviet citizens, and only one, standing on the sidelines, was, as it turned out later, a German flying to Berlin. As soon as we climbed the airplane ladder, we felt that we were familiar, bound together for a certain period of time by a single destiny; we were all equally interested in a successful flight. A morning haze hung over the city, and a fresh breeze walked around the airfield, but it was felt that the day would be hot again, as it had been the day before. We took our places without taking off our cloaks and coats. Soon the door slammed shut, the motors roared, and thoughts focused on the impending flight. The heavy craft ran along the concrete path for a long time, and when almost the entire airfield was already behind, it began to climb in jerks. "It is important to take off, but we'll definitely land somewhere," smiled one of the passengers. No one laughed, and, as it seemed to me, everyone found this joke inappropriate. I realized this and clearly embarrassed by the author of the joke.

Next to me was a man of medium height, blond with blue eyes, as it turned out, an employee of our trade mission in Paris, returning from vacation. We met. Dmitry Petrovich Gorbunov, looked closely, asked if I had to fly on this line, although judging by the awkward suit sitting on me, he was apparently convinced that this was my first trip abroad. And he was not mistaken. I had to go abroad as a cadet of a naval school or commander in warships, but then it was different, easier. So, in 1924 and 1925. our training ships Aurora and Komsomolets made their first overseas voyages, and we, cadets, not yet burdened with any serious responsibility, went through practice and saw only romance in these trips. These were small trips from Kronstadt around Scandinavia with calls to the Swedish and Norwegian ports: Gothenburg, Bergen and Trondheim. A few years later, I visited with ships of the Black Sea Fleet to Turkey, Greece and Italy. Not counting these short stays in foreign ports noteworthy, I replied that I had never flown abroad not only on this line, but generally never, ignoring caution for the sake of my service on warships.

A new acquaintance did not try to clarify why I was going to Paris, and I, of course, did not say about the ultimate goal of my trip. In the meantime, we climbed to the desired height and headed west.

The weather was great. Fields with ripe grain, meadows, forests were spread under the plane, and sometimes one could notice a straight line of the railway with a smoking steam locomotive and even a river. We flew at a low altitude. It was necessary to make a landing in Velikiye Luki, then to cross the Lithuanian border and continue on to Kovno, Koenigsberg, Berlin, Cologne and Paris. In those years, direct communications, as now, were few, and every two to three hours we landed, and even the planes on such a short route as Moscow - Paris changed two to three times. "In Velikiye Luki we can still have a good snack on Soviet money," Gorbunov, who had flown here more than once, initiated me when the plane made the last round before landing. Smooth landing, and we were already rolling along the grass field of the airfield. A few minutes later the plane stopped next to the white house of the airport. The small buffet was cozy, clean, and light snacks filled a small counter, behind which was a young girl who was welcoming the guests. Experienced passengers in Moscow already figured out how much Soviet money they would need here, and now they "walked the whole two-manded[11]", exchanging information, regarding who had how much left in their wallet after eating a couple of sandwiches and a glass of coffee or cocoa. But there was little time, and soon we again sat in our places and watched the airport disappear under the wing.

The pilot gained altitude, heading for Kovno, where we had to linger more than we should. Something happened to the plane, and we were offered to transfer to another machine, ready with a German crew. In an effort to maintain the schedule, and perhaps it was necessary, the pilot began to gain altitude without making circles over the airfield. I did not have time to take a closer look, as the city was already left behind, we were heading to the capital of Prussia - Koenigsberg. When the sea appeared to the right, experts informed that we would soon be landing. Under us was already fascist Germany or, more precisely, East Prussia, cut off under the Treaty of Versailles from the main territory of Germany by the Danzig corridor. Someone advised here to stick together and be more attentive. "Do you expect any trouble?" I asked Gorbunov. "I don't think so," he replied, "but special suspicion of Soviet passengers, and sometimes rudeness, does not exclude any incidence, and we don't need it." Smooth landing. The form of the airport employees resembled a military airfield. Abrupt command orders, a clear step and military greetings were then characteristic of all of Germany, and in particular to Prussia. Having first arrived on German soil, I looked closely at people and orders through the prism of the literature from the German literature about the Germans during the First World War, the revolutionary events after the surrender in 1918 and the onset of fascism in the early 30s. I recalled how a year ago, as the commander of a ship, I met the Bulgarian revolutionary G. Dimitrov, talked with him, admired his behavior at his trial, where he was accused of setting fire to the Reichstag.[12] Since then, fascism has become more active. Hitler was preparing for revenge.

---

[11] Two-manded is a Russian colloquialism for 20 kopecks coinage. Meaning here one knew exactly what to purchase with 20 kopeck coinage brought.

[12] Dimitrov was acquitted and expelled to the Soviet Union in 1933.

In those days, the role of the German command in preparing the rebellion in Spain was not yet clear to me. It was hard to predict that in a few months it would be possible for some of the German pilots at this airdrome to drop bombs on Cartagena or on ships that I could sail with the Spanish sailors. Leaving the plane, we stood near the airfield fence, waiting at the landing. Just a few minutes, and everyone was already following the tall reddish German to their plane. "Achtung - attention!" Someone shouted loudly, the motors roared, the plane took off from the ground. We flew over Germany. Unlike Lithuania, here the whole earth is divided into even squares or triangles; a dense network of highways and railways was easily visible from the plane. Passengers became more dominated by German speech. The stewardess served tea and cookies; two hours later we had a longer stop in Berlin. As you approach the capital of Germany, more and more roads, motorways converge at an preconfigured point. Landed at an airport, the name of which is not preserved in memory, somewhere between Berlin and Potsdam.

Positioned on the open veranda of the restaurant, we watched as one by one planes rose and landed. Suddenly the ambulance siren attracted the attention of all restaurant visitors. An unsuccessful landing, and the plane with its tail up lay or, more correctly, stood on one wing. Several cars were already hurrying to him. Soon life went on; obviously, the matter was limited to a small breakdown. "Donate to aviation," my roommate joked, although we ourselves had to fly out in a dozen minutes, and then flights were still not at all as ordinary and safe as they are today. In the afternoon we took off and headed for Cologne, saying goodbye to our two Soviet comrades who remained in Berlin. Sleepy. In Cologne, without leaving the plane, we waited the last stage of the flight - to Paris. Soon the plane took off. The weather turned a little poor, raindrops appeared on the glass of the plane, it somehow got dark right away, and when half an hour later they jumped out of the clouds, we saw a sea of lights. These were small glowing dots - they were visible right below the plane and on the horizon. Villages or large illuminated cities were observed on the right and left. Soon a bright stripe on the horizon, and a huge city was discernible - Paris, became a light show. On the map for passengers with a route laid on it, Orly airfield was designated as our final destination. Gorbunov tried to show me some notable places in the midst of this sea of lights, but, frankly, to no avail. Was it possible, for the first time being over Paris, to find the bright lights of the Eiffel Tower or the dark spots of the Champs Elysees? And thoughts were already occupied by another: who will meet me and how will it help me fly further, what will happen to me in Spain and when will I get there? "Your task is to arrive at the place as soon as possible and get acquainted with the situation in the Spanish Navy," they told me in Moscow. Now it was in fact to fulfill this instruction.

Soon the red lights on the ground marking the boundaries of the airfield became clearly visible - we were going to land. Everything that was observed from the air as something distant merged into one glow now disappeared from sight. Only a small, brightly lit area acquired a specific shape in the form of buildings with advertisements. Soon, the monotonous noise of the engines gave way to the dissonant sounds of human voices and the signals of cars. When I got out of the plane, I looked around, trying to find a comrade who was meeting me.

"Are you Kuznetsov?" A stranger of small stature asked me. "How did you find out that I am exactly who you need?" I asked A. I. Byaller, secretary of our air attaché in Paris, when we were already in the car. "Is it really difficult to recognize a person who has never worn a hat and a civilian suit, besides those hastily acquired in Moscow?" He smiled. To say that I never wore a civilian costume would be wrong; but it was so rare and long ago that I apparently, was different from all the truly civilian passengers. The money allocated for equipment in Paris was still in his pocket. It happened so often. It happened that our volunteer would come to Spain, would put it on himself in order to disguise himself as a Spaniard, and thinks that he can no longer be distinguished from any Andalusian, and he, like a white crow, stands out from those around him. But a few months will pass, he will pick up a hat for himself, suitable in color and size, so he will learn to wear it nicely on the back of his head or on his side, and it's really easy to take him for an indigenous resident of the Iberian Peninsula.

At such a late hour, and even on Sunday, it was difficult to find anyone at our embassy, and Byaller and I decided to go straight to the hotel. "I'm thinking of arranging you at the Saint-Germain hotel - it's very close to our embassy," my guide told me while we circled the unfamiliar streets of the mysterious city, which until now is known only to me from books and films. Not far from our embassy was a small hotel where many Soviet citizens staying, arriving for the first time from the Soviet Union to Paris. It is convenient, the person will not get lost and the very next day can independently get to the embassy. In addition, very close to the hotel was a small tavern-restaurant with a fat host and an even thicker hostess, who warmly welcomed all Russians and knew the tastes of our comrades.

Of course, I did not think or guess that the maid who met us, the Spanish Carmen, would later meet and escort more than a dozen of our volunteers and ask with partiality how the struggle was going on in her homeland. There were no such volunteers at that time, and she apparently didn't even have the thought to ask where I would go next. Yes, perhaps I wouldn't remember Carmen if I didn't take advantage of her kind consent to shelter me at this hotel a year later, when it was impossible to find a free room due to the world exhibition in Paris. "Take me to Saint-Germain," I told the taxi driver. Carmen asked me for a two-hour wait and, relocating someone, provided me with beautiful apartments. But we will return to this.

Byaller took me to the room he had prepared in advance. It was still not too late. What to do? I did not know how many days I would be able to spend in Paris and whether I would ever be able to return here. "If it's not a big deal for you, then take me to the city," I said to the very amiable Byaller. "Where?" He asked a question, and from his look I realized that the possibilities were great. This is Paris. I replied that I would like to have dinner and see the city. "Then to Montmartre," he snapped no doubt. The car was at the hotel. Byaller sat behind the wheel and, like an old Parisian, rushed off, deftly maneuvering among numerous taxis and walking public.

Here is the famous Montmartre. In the evening hours in Paris there is no other place filled with such a motley audience. Large and fashionable restaurants alternated with small cheap cafes with portable tables. Huge advertisements with colorful lights shouted visitors to cabarets and dance halls. We chose the middle between a luxurious restaurant and a modest cafe.

So that evening we saw Paris, which does not like to go to bed early and has fun almost until the morning.

I saw real Paris labor the next morning. From the hotel window, I watched as workers, employees of various institutions walked, traders rushed to the markets, and street cleaners swept the streets.

I had to urgently come to the Soviet ambassador in France, V.P. Potemkin. That's exactly what I did. Vladimir Petrovich immediately invited me to his place to briefly inform about the situation.

In more detail and from the point of view of military operations, we examined the situation on the Iberian Peninsula with our air attaché N. N. Vasilchenko, who also combined the post of naval representative in France. I wanted to learn as much as possible about the latest events in Spain. Byaller, the secretary of the air attaché was on duty perusing the Paris press. There was nothing comforting about it. The civil war took a protracted character. Despite the first successes of the republican government and the fact that the entire fleet and a few aircraft were against his side, Franco managed to gain a foothold on the peninsula and go on the offensive. The Germans and Italians openly sided with the rebels. Despite the bold active actions of the republican fleet, they failed to effectively block the ports of Gibraltar. Italian planes ensured the transfer of units from Morocco and reinforced the rebel garrisons. Badajoz, located on the Madrid-Lisbon highway, was captured by Franco in mid-August, and the northern group of rebel forces merged with the southern. The cities of Cordoba and Oviedo remained in the hands of the rebels, although it was announced on the radio that they had been freed by the Republicans. The Alcázar, a palace in Toledo where a handful of rebels stayed in the city center not far from Madrid, could not be captured.

From the newspapers one could understand that in the early days of the rebellion, Franco was unable to seize power, but the government also lacked organized military force and determination to quickly suppress the rebellion. Franco was supported by fascist Germany and Italy, while the legitimate government of the Spanish Republic did not receive real help from Western states, and platonic sympathy was not enough. That was the situation.

The next day, B. F. Sveshnikov flew to Paris. He was appointed as air attaché to Spain, and he, like me, had to get to the place of his new service by any means. We met with L. Ya. Gaykis, an advisor to our embassy who was entrusted with assisting us. Gaykis told how hard it was to find a plane to ensure the flight to Spain of our ambassador M.I. Rosenberg and military attaché V.E. Gorev.

"In a day, a Spanish military Douglas is scheduled to fly in to take spare parts purchased for republican planes, it should be able to take you as well," Gaykis said, but added that everything will depend on the amount of cargo.

"Well, we are not ambassadors, and, like military people, we may well be content with a modest place somewhere on the crates, if only the plane was not overloaded with more valuable details," Sveshnikov joked. These words were prophetic. We really had to settle among the boxes with spare parts, throwing out a few seats to lighten the aircraft. Since As far as the air route was concerned, I relied entirely on Sveshnikov. He knew French better than me, and he had a common language with the pilots. Just as the pilots who first enter the ship behave timidly and obediently, so I, the sailor, was willing to meekly submit to fate in an element alien to me - in the air.

But the "Douglas" expected from Madrid did not yet exist, nor was there any information about his arrival. Despite the military situation in Spain, we were still an attaché and thought a bit about the etiquette awaiting us. With the help of the energetic Byaller, we ordered state-owned suits from the Old England store. The girl who spoke Russian kindly served us and promised to make the first fitting almost the same day, and with a small additional payment, by the evening of the next day, give us ready-made suits. We even ordered striped trousers and bought fashionable hats that we never needed. In Moscow, I consulted on a military uniform, and decided that in the near future it would hardly be needed. And by the time it comes to formal dinners and receptions, a military suit can be ordered there or sent. In Paris, it was decided to stock up on only civilian clothes.

The day of a possible departure was already twice confirmed, but the Spanish plane did not arrive. I got used to it and looked through French newspapers myself in search of new information from Spain. I got a Spanish textbook and a small dictionary.

There was enough time, and B. F. Sveshnikov and I managed to explore some of the sights of the city. We climbed the Eiffel Tower, examined the Arc de Triomphe, visited the Tomb of the Unknown Soldier. We saw the Les Invalides, where Napoleon is buried, and visited the National Museum.

The famous pilot A.V. Yumashev[13] flew to Paris these days. He was preparing for a flight over the North Pole, and some formalities with records led him to the capital of France. On Sunday, Byaller organized an excursion to the Louvre for us - not taking advantage of this would be unforgivable. Let's go and not regret it.

---

[13] Yumashev participated in several Soviet sponsored aeronautical record making events.

At the Distant Meridian

In Paris, I had a meeting and first acquaintance with A. A. Ignatiev. He was then going to return to the Soviet Union, but for now he was engaged in the implementation of certain instructions of the trade mission. Byaller introduced us. Alexey Alekseevich was in a hurry; he had a briefcase in his hands. It turns out that in the brief case of the former tsarist military attaché in Paris were dolls, yes, dolls, one of which he showed us. He was going to show them to some company. At first, I did not attach any importance to this meeting, and only when we parted did Byaller talk in more detail about this man. Perhaps this episode would not be remembered, but after three or four years, fate again brought me together with A. A. Ignatiev. At one of the receptions in Moscow, he came up to me and introduced himself in a military fashion: "Lieutenant General Ignatiev." He was already in the Soviet Union and soon became famous for his book "50 Years in Service". I was not familiar with him close, but occasionally we met.

Returning to the hotel in the evening, we learned that the Spanish plane had finally flown in and was ready to take us if we did not claim any great comfort. We, of course, agreed and started packing. All thoughts from this moment have focused on Spain.

From Paris to Madrid

My participation in the Spanish war actually began in Paris. At Orly's airport, crowds of people cheerfully welcomed the appearance of a plane with Spanish decals on the taxiway. Landing began. There were few passengers: a few Spaniards who were carrying to Madrid spare parts and parts they had bought in France for the French Potez aircraft, which were then still in service with the Spanish military. Among the Spaniards was a representative of the Republican Ministry of Finance X. Lopez, with whom I subsequently was to meet. Having settled down on one of the boxes with spare parts during the flight from Paris to Toulouse, where we were supposed to spend the night, I tried to make acquaintance with my fellow travelers. Although I still did not know Spanish at all and did not speak enough French, with the help of gestures we quite successfully explained. The Spaniards turned out to be very friendly comrades.

Toulouse is a southern French city. It was to receive many volunteers later who made their way to the Iberian Peninsula in various ways. Toulouse hosted many major statesmen during the years of the struggle of the Spanish people. Some flew to Madrid or Valencia as well-wishers, others as observers. At the end of the war, Toulouse became a refugee camp from Spain, which the French government kept behind barbed wire to please Franco.

All this was still unknown to us, and B. Sveshnikov and I, settling in a hotel, went to explore the city. We went to a restaurant. An orchestra of "Russian Cossacks" played. We sent a note and watched as the Cossacks inspected the hall with curiosity, trying to guess which of the visitors ordered the "Club"[14].

Before going on vacation, the Spanish pilots gathered and began to discuss the route of tomorrow's flight from Toulouse to Madrid. They argued about how best to fly: through Zaragoza, which was already in the hands of the rebels, or turn east, clinging to Barcelona. We decided to fly straight. Early in the morning, when we arrived at the airfield, the technicians were already warming up the engine, and black haze along with flames and roaring flew out of the exhaust pipes. The cloud cover was so low that it was barely possible to see the roofs of the hangars at the edge of the airfield. At the appointed hour, we took off and, barely lifted off the ground, as a dense mass of clouds hid everything that was below. Ahead were mountains, followed by Spain.

The plane flew at an altitude of about 5 thousand meters. This was necessary for security reasons. They were approaching Zaragoza. The height of the aircraft was the only protection against anti-aircraft guns and fighter attacks. True, the probability of meeting with the enemy was not high in those days, and the crew headed straight for Madrid, over the territory of the rebels. I wanted to see with my own eyes, at least from a great

[14] A popular revolutionary song of early 20th Century Russia.

height, the country where I had to work as an attaché "in unusual conditions". Already the first meetings with the Spanish comrades showed that our role would not be to observe etiquette and ceremonies, but as provided for in the agreement with the Spanish government, active participation in the struggle of the Republicans against the rebels. But so far everything was unclear.

My new Spanish friends, who had already flown this route many times as if on a normal flight, continued the conversation they had begun about how the republican government would soon end Franco and the people would deal with the Nazis. I did not take my eyes off the window, hoping to see at least one Spanish city. The green areas of the eastern part of Spain were gradually replaced by gray fields scorched by the sun, and the further we went down to the south, the more monotonous and dull the landscape became. Flying over the area of Zaragoza, we wanted to see the front lines, but from such a height it was unthinkable.

"This is where we are," said the navigator who came to us and pointed to the route map. Zaragoza was already behind. We passed more than the half way point. With a gesture, the navigator showed that we were descending, a little noise in the ears. Everyone's mood improved noticeably, conversations became livelier. I remembered reading somewhere about the particularly blue sky in the Madrid area and looked up. The sun had already risen high, and the sky without a single cloud was actually blue. Later, in Cartagena, we loved the sky covered with clouds more, and it was even better when it was raining - then the appearance of enemy aircraft was ruled out. Unfortunately, there this rarely happened. To the right, in the distance, the mountains of Guadalajara were already visible - it means soon Madrid.

Since the plane flew in from Paris, and Soviet citizens were least expected, we were mistaken for Frenchmen. It took a certain amount of time until the aerodrome authorities inquired about the location of the Soviet embassy. The Spanish comrades who were flying with us from the Ministry of Finance kindly escorted us in their car to the Alfonso Hotel and said goodbye. It seemed that there was little chance of ever meeting them in the future, but our paths once again converged in Cartagena, and one of them I escorted to the Soviet Union on our transport *Neva*. He carried out a responsible assignment, which I will talk about later.

Approaching Madrid, they felt that life in the city was subject to extraordinary events that affected all sectors of the population. I happened to be in the southern and eastern cities: Istanbul and Port Said, Athens and Jaffa, I saw noisy streets there, not to mention the bazaars. In Madrid, I expected to meet the same picture. However, something else caught my eye. The streets of the Spanish capital were filled with people and cars. Here, next to the usual peaceful working life with brisk trade in shops and restaurants, military life went on.

People with posters and banners, with weapons in their hands, walked in a noisy crowd and, it seemed, some tried to shout out the others. Hundreds of cars with painted sides moved along the same streets. The Spanish driver does not tolerate low speeds, even in calm peacetime, and even more so considers it shameful for himself not to rush in such a fateful hour for his homeland. In addition, the city streets of Spain are not designed for heavy traffic: they are usually narrow and filled with pedestrians. Beeps often do not help, and then the driver tries to compensate for this with his voice, and sometimes with threatening gestures.

On the fifth floor of the Alfonso Hotel, where our embassy was located, it was empty. The Soviet colony was still small, and B. Sveshnikov and I were given the choice of occupying any free rooms. From the street came the voices of people, the sounds of cars and individual gun shots. The location for the embassy on Alcala Street was hastily chosen and clearly unsuccessful. It is no coincidence that after a week or two we all moved to the Palace Hotel, located in a quieter area.

A few hours later B. Sveshnikov came for me, and together we went to report to Ambassador M. I. Rosenberg about our arrival.

Having dealt with how we flew, our new plenipotentiary immediately got down to business. He briefly outlined the political and military situation, emphasizing that he needed our help in order to have a clear idea of the republican fleet and aviation.

"You will learn the details from Gorev," he concluded the conversation and asked, without wasting time, to get down to business. He was extremely busy and was in a hurry to leave.

Like the ambassador, one of the hotel rooms was reserved for V.E. Gorev's office, where a large table was set, on which maps of the fronts already lay. The first time I saw Vladimir Efimovich Gorev. Dressed in a civilian suit, he remained a military man, fit, taciturn and able to summarize his thoughts. There are people who from the first conversation are able to make subordinates respect them and be afraid. As a military attaché, I considered Gorev senior, I expected that he would have to work a lot, and his attitude to the fleet was not indifferent to me.

"Spain, as you see, is surrounded by seas, and therefore the fleet can play an important role, can it not?" Gorev asked me when we met to outline a plan for my work. The circumstances were such that it was my only professional conversation with him. At the first opportunity, I left Madrid for the Spanish sailors, and he plunged into practical work on the defense of Madrid. As the chief naval adviser, I had more contact with senior combined-arms adviser Y. K. Berzin and the ambassador. Moreover, in November they were already in Valencia, and Gorev remained in Madrid until the end of his work in Spain. This did not stop us from meeting and discussing common topics several times: he shared information about the situation on the Central Front, and I told him what the republican fleet was doing. I recalled Gorev and at home G. M. Stern spoke very well of him. "You rarely see such a brave man," he said and gave examples of Gorev's surprisingly bold behavior during the days of the defense of Madrid.

A year later in Vladivostok where I commanded the fleet, I became closely acquainted with V.K. Blyukher. In the first conversation, the marshal, being interested in affairs in Spain, in particular asked if I had met with Gorev. It seems to me that they were still familiar through collaboration in China. He also praised Gorev as a very knowledgeable commander. A little later, a sad fate befell both of them: both were repressed.[15]

But back to our first meeting. "We will have to fulfill our duties as an attachés in a difficult environment, which I'll try to tell you about now," Gorev began, inviting us to sit down and follow his story on the map. "The decisive front in the near future is the central one, where there is a struggle for Madrid," he said and pointed to the map. It showed how a few weeks ago the northern and southern rebel groups were separated and the city of Badajoz on the Madrid-Lisbon highway was in the hands of the Republicans. But the fresh lines of the blue-red pencil marked the new position of the struggling sides. Badajoz fell, and the rebels were able to direct all efforts to attack the capital. The day before the fascists occupied the city of Oropesa and are now advancing towards Talavera, a small city, but from a military point of view, a very important point, located on the way to Madrid. This, according to Gorev, is very worrying for the republican government, although in the north-west of Madrid the rebels were even closer, but there the terrain made it possible to create a strong defense. In the south of the country there were heavy fighting for Cordoba. The northern provinces of Spain - Asturias and the Basque Country - fought heroically. The latest news from there was disappointing: the rebels attacked Irun, and communication with France was already cut off, and the surrounded Oviedo remained in Franco's hands. Despite this, the vast territory in the east of the country was under the control of the republican government. There was still hope for an early suppression of the rebellion. We agreed to postpone a more detailed discussion of the situation at sea the next day.

"Try to get to know the situation faster," Gorev said farewell, confirming the next hour for, "purely naval," discussion.

---

[15] Executed. Victims of the Stalinist purges of the office corps.

It was impossible to get acquainted with the situation during the day, but it turned out to be quite useful to gather thoughts and look through the latest French newspapers. Over the past week, while still in Paris, I tried to find all the notes concerning the republican fleet in order to have an idea of the actual state of affairs. I already knew that the main core of the Spanish fleet remained loyal to the government and two or three rebel ships, including one battleship "Espana" and the cruiser "Cervera", were already operating in northern Spain. The information regarding the new cruiser "Canarias" was contradictory: some newspapers claimed that she went to sea manned by German petty officers, while others considered this unlikely.

Much had been written about the actions of the Republican fleet in the Gibraltar area, the landing on the island of Mallorca and the alleged capture of Algeciras, located in Gibraltar, and even the port in Morocco - Ceuta.

Having collected all these data, I prepared after reporting to Gorev to leave for Malaga or Cartagena as soon as possible. But how to do this I still had no idea. The official position of the naval attaché obliged me to pay a visit to someone in the Ministry of the Navy and get my consent to travel there. M. I. Rosenberg promised to help, but he himself still did not have time to contact all officials. I decided to consult with V. E. Gorev. He was already with the Minister of War Sarabia, but did not know at all what was being done at the Ministry of the Sea. And in the navy ministry, luxuriously furnished like all the ministries of Spain, there was no master yet. Socialist Indalecio Prieto claimed the post of Minister of the Sea. Despite his open dislike for Largo Caballero, he supported his candidacy to replace Giral. Engaged in the unfolding political games, Prieto has so far been calling up newspapers to fight the rebels. I did not want to wait until Prieto officially assumed the post of minister, and I decided to get an appointment in the coming days. Having reported to Gorev my meager information about the Spanish Republican fleet, I turned to talking about my practical activities. It was decided that I had nothing to do in Madrid. In the current situation, the naval attaché should provide real help, and this can be done where the republican squadron is based and operates.

"I will try to talk with the ambassador on how to expedite your visit to the Ministry of the Sea, and you, without wasting time, are to look for ways to establish communication with the sailors," Gorev concluded our conversation.

Several days wait in anticipation of the opportunity to meet with I. Prieto were not in vain. I made useful contacts with people and looked around the city. One evening I met with M. Koltsov. The small room, adapted for the dining room, was crammed. For dinner, embassy workers living in the Alfonso Hotel and several guests were gathered. Among them were M. Koltsov and cameraman R. Carmen. They had just returned from Toledo, where they watched an unsuccessful attack on Alcazar Castle. Koltsov vividly described instead of a decisive attack, negotiations were going on between the Republicans and the rebels, if necessary with the help of guns, who settled in a building located on a mountain in the center of the city of Toledo. Later, I managed to visit there before Franco's units captured Toledo. By this time, the fall of Talavera was confirmed, and the Republicans frantically assembled, armed and trained militia units to block the enemy's path to the capital. They did not speak of an imminent victory over the fascists; it became more apparent every day that the civil war was becoming protracted. At the end of the dinner, I had a conversation with Koltsov. The reason for him was his desire to learn the latest news from Moscow: he had left there two weeks earlier. I spoke about the great interest in Spanish events in the USSR, about the sympathy with which everyone relates to the struggle of republican Spain. I was not talking about Moscow, which was only one day, but more about Sevastopol, about the sailors of the Black Sea Fleet.

"My brother three years ago made a trip on some cruiser to Greece and Italy," said Koltsov. It was about the artist Boris Efimov. At the end of 1933, a detachment of our ships really sailed to Naples. This was a return visit to Batumi with two Italian submarines. I was then a senior assistant[16] on the cruiser "Chervona Ukraine" and remembered well that the famous artist B. Efimov and the writers I. Ilf and E. Petrov took part in this campaign.

We sat for two hours. I willingly recalled the campaign and meeting with B. Efimov. Koltsov shared Spanish news. Mikhail Efimovich already managed to visit Barcelona, visit the central and Aragonese fronts, met with many political figures: with Giral, Caballero and others. He was already received by President M. Azana and met with the leaders of the Spanish Communist Party X. Dias, D. Ibarruri and V. Uribe. He was going to visit republican ships and even take part in one of the military operations.

When asked how he assesses the situation, Koltsov replied that Franco had organized strength and military knowledge, which the Republicans lacked, but he believed that the people would find the strength to defeat the rebels. Saying goodbye, we hoped that we would meet often, but in fact, after that I only saw Koltsov in Barcelona when I spent several days there with Minister of the Sea Prieto in April next year.

In the same days I met M. Pasqua. He often went to the hotel, kindly offered his services as a guide. B. Sveshnikov and I more than once took advantage of his offer without thinking about his official position. "You are not particularly familiar with Pasqua, because he is already the Spanish ambassador in Moscow," one of the embassy workers told us once. A few days later, he really left for the Soviet Union, and, meeting his last name in the newspapers, I, embarrassed, recalled how I bought beret at the store with his help. Later, I happened to see M. Pasqua in the summer of 1937. He came to Spain and went to the Metropol with our ambassador. M. Pasqua was the first ambassador of Spain in Moscow. As you know, the mutual recognition of the Soviet Union and Spain occurred in 1931, after the overthrow of the king and the proclamation of the Republic. However, the states did not have time to exchange ambassadors, as other events had come -

[16] Soviet naval rank of the era equivalent to Lieutenant in the USN/RN.

the elections to the Spanish Cortes in 1933, which brought right-wing bourgeois parties to power. Julio Alvarez del Vayo, already appointed ambassador to the Soviet Union, did not travel to Moscow, and the Soviet ambassador to Spain, Anatoly Vasilievich Lunacharsky, was forced to stay in Paris, where, unfortunately, he fell ill and died. In a short conversation, Pasqua spoke with admiration about Moscow and our people. His further fate is unknown to me.[17]

A few days later, when I was already going to Cartagena, a tall man came up to me and asked in French when I was thinking of going back to Madrid. It was Andre Marty. His name was known in the Soviet Union in connection with the uprising on French ships in Odessa in 1918, and in Sevastopol even one destroyer was named Marty. It turns out that he arrived in Spain to take part in the struggle, and according to old memory was interested in the actions of the fleet. He asked to be informed about the affairs of the Republican ships when I returned to Madrid. However, Marty soon left for Albacete, where the first international brigade was being formed, and I never saw him again.

Using my free time, I acquired textbooks of the Spanish language and daily learned a few common phrases or words, immediately practiced their practical application. My idea of the affinity of French and Spanish turned out to be erroneous. The Spaniards did not understand me, and I was helpless. The embassy already had several translators, or rather translators who worked very well with our people in the rear and on the fronts. I had to work on ships, go to sea, which is unacceptable for a female translator. I know that throughout the war in Spain they worked in the most difficult conditions and were not inferior to men in courage, but they were not in Cartagena or on the ships. I had to rely only on myself until with the arrival of our volunteer sailors several male translators were sent to us. It also became obvious that for serious and long-term work in a foreign country it is necessary to start with studying the language otherwise you will not learn much or will be bound to a translator. And God forbid, if

---

[17] Marcelino Pascua survived the war.

the translator turns out to be an active comrade and wants to not only translate, but also express his opinions during the negotiations. Therefore, N.P. Anin who arrived in mid-September and did not speak Spanish, I recommended to quickly learn the language.

By the way, during the campaign, Anin was on the flagship destroyer "Lepanto" together with the flotilla commander B. Ramirez. Already at sea, I saw how necessary it is to have constant contact with him. I sent him a semaphore message written in Latin script, but using Russian words, and only at the very beginning I wrote in Spanish: "Companero Ruso" so that the semaphore gets to its destination. The experience was a success: during the entire voyage from Cartagena to Cantabrico, we perfectly conveyed the necessary messages to each other. In the months that followed, when our volunteer seamen numbered several dozen, all those who had studied Spanish assiduously were noticeably better at their work and more useful. But these were later times. Our volunteer comrade, K., arrived to assist me and I had to fly somewhere for ten days, and he alone remained in Cartagena. Having placed him in my room, I gave him textbooks and asked, without wasting time, to start learning the language, and jokingly added that he would remember the word “komida”, that is, lunch, and this would already be a guarantee against starvation. Returning to the base, I was genuinely indignant when, when I asked how the matter with the language had progressed, he replied that he had remembered the word “komida” and had used it very well in my absence. He was a good friend, but his reluctance to learn the language a few months later led to the need to send him back to his homeland: you can’t advise anything to Spanish comrades on the ship, knowing only two words: No - no or Si - yes. And so it was. This is not an idle question. True, as a rule, everyone arriving in Spain quickly mastered the necessary vocabulary and successfully completed their tasks helping their Spanish comrades as best they could. As a pleasant exception, I want to mention C. Ramishvili, who arrived in Spain with a well-known knowledge of the French language, quickly learned Spanish and soon without the help of a translator performed very complex advisory functions with the commander of the Cartagena base A. Ruiz. Ramishvili not only easily spoke with the Spaniards but learned a lot of Spanish sayings and was free to conduct a conversation on any topic. When asked how he spoke the language, Ramishvsch jokingly answered: "Perfectly like Cervantes" - that is, no worse than Cervantes himself. However, all this wass already much later. So far, unfavorable news has come from the fronts every day.

Although the rebel offensive from the north was suspended in the mountains of the Sierra de Guadarrama, a new threat from Extremadura had already been identified. The African units of the rebels were rapidly advancing towards Madrid. The day before was the first bombing of the capital. Bombs fell on the outskirts without causing harm, but there were a lot of firing, or rather, shooting. The entire armed population, when enemy aircraft appeared, randomly started firing pistols and machine guns. They warned that it was dangerous to stick out of the windows, since you could become a victim of a stray bullet. As regards the fleet, it was only possible to find out that it is based in Cartagena and Malaga, operates somewhere near Gibraltar and seems to be preparing for some major operation. All this is rather vague, but no one else could say.

From day to day, they have been waiting for big changes in the government. The left-wing government of Giral, which has been in power since Franco's rebellion, was unable to wage an anti-fascist war and was living out its last days. Inevitable was the formation of a stronger government with the participation of socialists and communists. This has already been negotiated. No one knew which member of the government now ran the fleet. It was only known that the socialist Indalecio Prieto in the future coalition government is likely to take the post of Minister of the Sea, but for now ...

While B. Sveshnikov and I met within the city, again drew attention to the animation on the streets, to a lot of flags and posters. Without exception, even young girls were dressed in dark blue overalls with a zipper and did not wear hats. Plain clothes with a tie and a hat were clearly out of fashion. Many walked with weapons in their hands. To my remark about the big movement, someone replied that it was nothing compared to Barcelona.

Naturally, I wanted to understand at least a little the difficult political situation that existed at the end of August 1936. The position of the naval attaché obliged me to do this, and besides, I remembered the instruction of my Moscow authorities that without understanding the political I am not guaranteed against committing major mistakes as a representative of the Soviet Union. Mikhail Yefimovich Koltsov and those comrades from the embassy who had already managed to see and evaluate the domestic political situation with their own eyes told me a lot. The Republicans were in power, led by President M. Azana and Prime Minister H. Giral. Both of them, no doubt, were opponents of fascism, were kind to the restoration of diplomatic relations with the Soviet Union and were grateful for the sympathy and help that our people had already begun to render to struggling Spain. At the same time Republicans very hesitantly and cautiously approached the problems that were ripe within the country, promising a gradual and consistent elimination of all vestiges of feudalism and the implementation of the necessary reforms. The most pressing issues it seemed to me, were agrarian reform and the national question, especially in Basque country and Catalonia. In practical activities, President Azana and Prime Minister Giral turned out to be poorly adapted to governing the country during the days of the civil war. The government was clearly unprepared for a quick and decisive struggle against the fascists, although the uprising was brewing almost openly and it was necessary to think much earlier about the possibility of a difficult situation. Before the rebellion, the Communists were accused of seeing fascism everywhere, and when the rebellion became a fact, the ruling party released the reins of government from its hands and was at a loss. In opposition to the ruling republican party from the first days of the rebellion stood the left faction of the socialist party, whose leader was L. Caballero. At the moment described, the disagreements were already so sharp that Caballero, according to Koltsov, said without hesitation: “This is a comedy, not a government! This is a shame! ”But the position of the socialist party was far from monolithic. The party leadership and I. Prieto disagreed with L. Caballero on many fundamental issues of warfare and domestic politics. Prieto did not hide his attitude to Caballero and awarded him unflattering epithets.

The socialists did not have a common opinion even on the most important issues of the struggle against fascism. It should be noted that Prieto in some matters better understood the situation. It was he who, after the fall of Malaga, when the communists decided to leave the government, said: "Without the communists there can be no government."After this, Caballero left, transferring the post of prime minister to H. Negrin. But more about that later. The situation was different in the Spanish Communist Party. The inter-party intrigues were alien to them and their purpose was clear. The party had strong leaders: X. Diaz, D. Ibarruri. V. Uribe, N. Cheka, A. Mikhe and others. But the party was young and only gained influence among the broad masses of the people. Their efforts were concentrated mainly on the central front.

A considerable and often negative role was played by the Spanish anarchists. From leftist phrases in the rear, they avoided going to action at the front. Their inconsistency and a peculiar understanding of the role of discipline brought great harm. True, the danger united all anti-fascist forces, but it was impossible to talk about a united and firm leadership. Replacing the government of Giral with a coalition government with Largo Caballero at the head did not give a significant improvement, although it was a step forward. Now we have to recall with bitterness how often the enthusiasm of the people was not used due to the lack of a clear goal and firm leadership. Examples of the suppression of the rebellion in Madrid, Barcelona, the Navy and in the famous Montaña barracks in the very center of the capital, serve as proof that, if the Spanish people had firm leadership, the rebels would not have lasted even two weeks.[18]

Martial law in such a political environment was unclear. The rebels for a long time preparing for the fight and developed detailed plans. They enlisted the support of Germany and Italy in advance, not to mention Portugal. England or France helped them willingly or involuntarily, hiding behind the slogans of "non-intervention". The time factor in fast-paced events was of great importance. Franco's supporters provided him with immediate and real help; an example of this was the transfer of Moroccan units to the peninsula on Italian military aircraft. As for the republican government, practically only the Soviet Union supported it in almost three years of unequal struggle.

I got to the Ministry of the Navy a few days later. At the appointed hour the embassy car drove up to the Ministry of the Sea. Apparently warned, a guard of sailors greeted us at the entrance to the building. All ministries in Spain, according to experts, are luxuriously furnished, and the military in particular. Already in the lobby there are many paintings, carpets and upholstered furniture. In the huge office, resembling a rich boudoir rather than an office, sat the future Minister of the Sea Indalecio Prieto. Not very tall, very thick, with small, floating eyes, at first glance he seems clumsy and lazy. But it is worth meeting his gaze and you will see that he has a lot of energy and experience. According to Spanish tradition, he was often called by the name: Don Indalecio or just Indus. He enjoyed great authority. For more than 30 years he has been involved in politics and elected to the Spanish Cortes since 1918. Now he is not yet Minister of Marine, but he has already agreed to this portfolio in the government of L. Caballero. When we entered, he cordially met our Ambassador Rosenberg and invited us to sit down. "I am not yet a minister to make decisions on the fleet, but now I will introduce you to one of the members of the Central Fleet Committee and I will not mind if you go with him to Cartagena or

---

[18] Regular Army uprising in Madrid defeated by Civil Guards and hastily armed civilian volunteers during July 18-20, 1936.

Malaga," said I. Prieto in response to my desire to meet in place with the republican squadron. During the conversation, I looked closely at him: after all, I had to meet with him often, and my work largely depended on his decisions. Indeed, I got to know Prieto more closely, when for many months the republican fleet was primarily engaged in ensuring the delivery of weapons from the Soviet Union and my responsibilities included arranging the meeting of each transport far out to sea and ensuring its safe arrival in Cartagena. I often needed the help of Minister of Marine I. Prieto. I will talk about this peculiar figure a little later, but now I was only interested in the opportunity to go to the fleet's bases. The sailor Prieto introduced me to was Pedro Prado, an active organizer of the suppression of the rebellion on ships, who was then selected as a member of the Central Fleet Committee, in the hands of which was in fact all power, since the fleet had not yet been controlled from Madrid.

M.I. Rosenberg stayed with Prieto, while Prado and I left, having received permission to leave for the fleet together. Prado was constantly on warships and did not have a permanent desk in the ministry. Pedro invited me to a bar located on the lower floor, and ordered a couple of bottles of cerveza - beer. He was fluent in French and we easily agreed on the time of our departure. Prado was to my delight a communist and perhaps the only one in a leading position in the current fleet. In the Republican Navy, during the years of the Civil War there was a struggle for influence mainly between Republicans and Socialists. Prieto, becoming a minister, took all measures to ensure that the key positions in the fleet were taken by his supporters - socialists.

Prado - tall and thin in a gray navy mono - smoked one cigarette after another. Like all southern Spaniards, he spoke quickly, swallowing individual sounds; he sincerely wanted to tell me about everything that was happening and he told in detail about how the rebellion occurred on the ships and what the squadron did in the fight against the rebels. However, there was little time, and we soon parted, agreeing to meet the next day in order to go by train to Cartagena. "Hasta la Vista" (goodbye), my new acquaintance said goodbye, taking me to the car. This was at the end of August 1936. I began to prepare for the first trip to the ships of the Spanish fleet.

In a new suit bought in Paris I felt too elegant for the surrounding military situation. It was necessary to have something simpler. Pedro, smiling, said that until Cartagena my attire does not matter much and there he will provide me with a naval mono and the most fashionable beret. On the appointed day and hour, Pedro drove me to the hotel and together we went to the station. The train departed from Madrid in the evening and arrived in Cartagena in the morning of the next day. When we left it was not too late and my companion continued his story about the actions of the republican fleet during the rebellion. The map spread out on the table made it easy for us to understand each other.

Cartagena

Early in the morning, the train stopped near the beautiful station of Murcia. This is the capital of the province, which includes Cartagena. In Murcia there was a significant influence of the anarchists and the city was still troubled. According to Prado, the night skirmishes are still ongoing, which we saw very soon.

The train crossed the last mountain range in front of Cartagena. The sun had already risen high, and even through the window of the car one could feel how it was baking. But here is Cartagena.

Upon arrival, I stopped at the base commander Antonio Ruiz. Prado, busy with his own affairs, left me in Cartagena, and the next day he flew to Malaga. Since the Ministry of Marine has not yet formalized my permanent stay in the Navy I could not take an active part in any operation and I still felt unprepared enough for this. I had several days at my disposal and I could get acquainted with the city, its port, arsenal, docks and workshops. I learned a lot of interesting things.

Cartagena is a very old city in Spain, perhaps the oldest. In the suburbs you can find the remains of ancient canals and fortifications. “How long has Cartagena been around?” I once asked Antonio Ruiz. “The history of Spain begins with Cartagena,” he replied in jest, in which there was a lot of truth. After the first Punic War, when Hannibal took his campaign against Rome, he already used New Carthage (Cartagena) as a stronghold in the southeast of the Iberian Peninsula. It was not only a fortified city, but the center of Carthaginian possessions in the area. In 218 BC troops marched through Cartagena to cross the Alps and defeat the Romans at Cannes. It seems that only Malaga can be compared in age with Cartagena. The first colonies were founded in Cartagena, first belonging to the Phoenicians, and then to the Greeks. Cartagena has the right to be considered one of the oldest coastal cities in Spain, and as a port and fortress it is by far the oldest.

For more than twenty centuries, Cartagena has played an important role, being the best natural harbor on the Mediterranean coast of the Iberian Peninsula. Situated in the depths and sheltered from all sides from the winds, the bay around which Cartagena grew up has always been an almost ideal shelter for merchant and combat ships. High mountains, from all sides guard the bay, even today are of significant importance from the point of view of defense. The artificial ports built in recent decades - from Malaga to Barcelona - are inferior to Cartagena, primarily militarily.

Since the end of the XIX century Cartagena was in decline as the Spanish governments of this period supported El Ferrol on the shores of the Atlantic Ocean as the main naval base of Spain. As a result Cartagena faded into the background. Narrow, crooked streets and ancient buildings with thick walls, gray with time, heat and dust are characteristic of Cartagena. This is a fortified city in which it is easier to find traces of antiquity than the various improvements of today. In Madrid or Barcelona you can see tall modern houses, metro, wide promenades, luxurious boulevards, and fashionable hotels. None of this was in Cartagena. Here many streets are so crowded that it is impossible to drive by car and

God forbid meeting on the streets of Cartagena another car or a donkey pulling a huge cart. "Burro!" (Ass) (although this was related to the driver) - in this case my driver Ricardo shouted angrily, if I ran into resistance I took up arms. But it was worth resolving the dispute peacefully, and the opponents - such is the Spanish character - parted friends. "Salud," one will say and raise a clenched fist. "Salud, companero," the other will reply, although these people did not know each other before.

Cartagena shocked me terribly with its sharp social contrasts. Here were luxury and poverty, splendor and dirt, comfortable Hispano-Suiza cars and small donkeys pulling heavy carts. The huge building - the residence of the base commander - on Calle Mayor Street, half of which was occupied by his private apartments with a church and a large garden, and for the large guard there is one small room where soldiers or sailors lived in the mud, having only a trestle bed and the invariable for-all-occasions of life blanket cloak. Beautiful cabins for officers on ships and completely unequipped tight berths for the rank and file. A large salary for officers and insignificant allowance with poor food for ordinary and non-commissioned officers.

The harbor area of Cartagena consisted of internal and external harbors. The inner, natural, harbor was small: only destroyers and submarines could be based in it. On the shores of the inner harbor were docks and a shipyard. The outer harbor, formed by an artificial pier, was extensive enough for anchoring cruisers and even a battleship, but the means for supporting large ships were not enough. At the beginning of the rebellion, the dock contained only destroyers, and fuel depots were limited and poorly sheltered.

To protect the base from the sea, batteries of large (up to 15 inches) caliber and a large number of smaller guns were installed at nearby height. Anti-aircraft guns were, however, weak, and only a few 3-inch anti-aircraft guns with obsolete fire control devices covered Cartagena. Near the city, two airfields were located - Los Alcazares and San Javier, on which fighters were based during the war to protect the main fleet base and port.

Apart from Mahon on the island of Menorca - a base located in the immediate vicinity of enemy-occupied points on the island of Mallorca, Cartagena was the only naval base of the Republic on the Mediterranean Sea. Numerous trading ports were able to fulfill an auxiliary role for the short-term basing of ships, but of course they could not provide their permanent basing and supply with everything necessary. However, it so happened that during the three years of struggle, sometimes with great difficulties, the squadron was based on Cartagena and the modest trading port received a large amount of military cargo for the front. The whole life of the city was subordinated to this.

The face of the base changed during the war. Rebel air raids forced a strengthening of anti-aircraft defense and increased aviation to cover the base and assist the fleet. Large cargo operations forced the port to develop, build shelters and equip the port with cranes. The shipyard was completely switched to ship repair and the existing dock for the destroyers after the explosion on the cruiser "Cervantes" was extended to the required size.

However, I run a little ahead. At the time of my arrival in the inner harbor which was called Arsenal there were several destroyers returning from operations off Gibraltar and taking ammunition to go out to sea again. In the dock was the destroyer "Antequera", repairing boilers after lengthy combat operations. The outer harbor was empty, as the battleship and cruisers were in Malaga and Almeria.

The ships that I saw in the Arsenal were dirty, which is probably acceptable after returning from the sea but the noise and jostling on the upper deck spoke of the lack of proper order although the sailors dressed in a work dress worked with enthusiasm.

Here I want to make a small digression.

Observing the energetically acting republican non-commissioned officers and sailors I repeatedly and then involuntarily wondered: what did they lack for the quick suppression of the rebellion? After all, these are people on whom the legitimate republican government could certainly reliably rely. I did not find the answer right away. This required months of joint combat work, several trips to the sea and direct contact with personnel in combat conditions.

A clear goal, enthusiasm and strong leadership are the conditions necessary to defeat the rebels. Enthusiasm and heroism were on the ships until the last days and the sinking of the "Baleares" in March 1938 serves as proof of this. The goal - the fight against the rebels and the preservation of the Republic - was the same for all political parties on ships - from anarchists to republicans, however, it was understood differently. The main thing that was missing: a unity of views on the path to achieving this goal. Strength of leadership during a rebellion requires more than ever. But this was not. The government of M. Azana before the rebellion did not believe in danger from fascism. The same opinion was shared by the short-term prime ministers of Spain during these stormy days - S. Quiroga and M. Barrio. During the days of the rebellion, Giral's government was perplexed and the fronts had no leadership. The situation has improved very little and Caballero, a trade unionist who denied the need for a regular people's army could not radically change the situation. Therefore, it is not surprising that the fleet for a long time was without proper guidance from the center. Excellent opportunities, say, to control Gibraltar[19] were missed: the fleet alone could not do this, and no one was engaged in its interaction with the army. The consequence of poor leadership was the lack of sufficient order in the fleet itself. The situation in the fleet was very difficult in the midst of mass treason on the part of the officers and as part of the process of developing the military organization in the Republic after the rebellion I trained bold new commanders whose rank promotion was delayed.

Left without officers, the overwhelming majority of whom escaped to Franco or were neutralized at the beginning of the rebellion, the personnel understood that everything must be done to protect

---

[19] Strait of Gibraltar is presumed as Gibraltar proper is a British possession and the Royal Navy possessed overwhelming strength.

the Republic from fascism. Sparing no strength, they worked and carried out the orders of the Central Committee and the committees on ships that became the head of the fleet. This was done with a sincere desire to defeat the enemy. Efforts are not spared. Unfortunately, these efforts were not properly organized. Having escaped from the yoke of officers, the sailors did not want to recognize the old stick discipline. That was understandable. But the fact that a new, conscious discipline was now required even more than before was not yet understood. It took a certain amount of time to realize that the sooner a new firm order is established, the better. Unfortunately, many of the remaining officers often looked at the rebellion and war as an event that did not affect the foundations of social, and even more so military order, they slowly adapted to the requirements of wartime.

The petty officers and sailors who came forward during the period of rebellion and war were much more conscious. They were ready to work day and night and looked at the battle with the fascists as a deadly struggle. They could quickly master a new duty and even replace officers who had turned away from the Republic. Indeed, in peacetime, it was they who actually carried on their shoulders the whole burden of naval service and control mechanisms. An example is the head of the Lago water area guard, who, with exceptional energy and integrity, did his duty. There were many such people, and from them it would be possible to quickly nominate and train suitable officers. However, the leaders of the republican government and the naval department underestimated the importance of such a move, and Minister of Marine I. Prieto clearly sabotaged the creation of a truly democratic naval officer corps.

All of the above in miniature was reflected in the destroyer "Antequera", on whose deck I boarded with Pedro Prado. The crew has more than enough desire to fight for the Republic, but they do not strike against the fascists with a big weighty fist, but with spread fingers.

Time passed, and much has changed in the fleet as a whole and on this ship in particular. When in April 1937 I was on it during one of the operations there were no committees. The commander of the destroyer flotilla V. Ramirez gave orders, and the commander answered: "Aye!" and carried them out.

The enemy aircraft almost never bothered Cartagena. There was only one raid two days before my arrival, during which a house was destroyed and several citizens were killed. , people crowded around the house during the day indignant at the barbarism of the fascists. "Caramba", "Hombre"[20] every now and then was heard from the crowd.

The war in those days was still quite strange. For a long time not seeing big wars, the Spaniards were amazed at the sight of one fallen bomb. A desperate and hard exhausting struggle was going on in some sections of the front – for example, near Madrid – and somewhere near Zaragoza the anarchists did not fight at all on Sundays and managed to exchange cigarettes and fruits with the Francoists on the occasion of the holiday. No one had yet imagined that the struggle would drag on for almost three years that the same Cartagena people would not be surprised at bombing throughout the night and dozens of destroyed houses, burning warehouses and numerous victims.

Although the sailors already knew German ships had helped Franco and prevented with maneuvers the republican squadron from firing at Ceuta and Algeciras; the German consulate was still in Cartagena and warships of Germany and Italy could be seen on the roads. All with obvious displeasure pointed to the German consul walking along the promenade from where it was possible to perfectly observe all the movements of ships, but diplomatic etiquette did not allow decisive measures to be taken against him: official relations between Berlin and Madrid were not yet broken.

Return to Madrid

[20] Author's note. Exclamations expressing surprise, indignation.

After a brief acquaintance with people, ships, and the base on September 3, I returned to Madrid. I had to report on the situation in the fleet and get permission from the Ministry of the Navy for a longer stay in Cartagena and on ships. The car quickly raced along the coastal road in Alicante. In addition to the driver, I was accompanied by two sailors whom I did not know at all.

Commander of the base Antonio Ruiz advised to take them "just in case." I could talk to them only on general topics and even that mainly with the help of gestures. My companions apparently wanted to tell me a lot, but making sure that I did not understand everything, I limited myself to subjects that could be seen or felt. As it turned out later, both "marineros" were anarchists, who for some reason were considered the most reliable for protection.

At the airfields of Los Alcazare and San Javier, which we drove past, there were several old bombers.

"Potes ..." "Frances," the driver told me and pointed to the planes.

No one knew then that in a few months at these airfields a lot of work would be done to assemble and test new fighters and bombers which arrived from the Soviet Union.

Here is Alicante. Not at all like Cartagena. A resort town with a wide seaside boulevard, an arc enveloping the bay. Instead of gray government buildings and scorched slopes of the mountains of Cartagena there is a lot of greenery, palm trees, bright houses. Many restaurants and cafes with tables right on the boulevard. Peaceful life and no sign of war. Many foreigners walking and luxuriously dressed ladies. Mono is not in such fashion here as in Madrid or Cartagena. The driver glances at passers-by and indiscriminately calls all fascists. In the roads are German and Italian warships.

In the evening we left on the road to Albacete and Madrid. Near Aranjuez, the driver, new to this road, mistakenly turned to Toledo instead of Madrid. After driving about five kilometers, we hit the tail of a column of cars going to the front. And only then learned that we were lost. The driver turns back, and one of the guards, waking up, shouts: "Burro!" (Ass) and grabs the gun. Then it was easy to pass the front line without stopping and be in the arms of the enemy.

In Madrid, a change of government took place these days: the left-republic government of Giral departed, the coalition government of Largo Caballero came in, which was attended by six socialists (including the prime minister), two communists, three republicans of different styles and one representative from Catalan and Basque nationalists. The new government received a heavy inheritance from its predecessor. The situation at the fronts worsened: the rebels captured Talavera (120 kilometers from Madrid). Franco officially announced that he was launching a general attack on the capital.

By that time, the Soviet Mission had moved from the Alfonso Hotel to the Palace Hotel. From our comrades I received useful information about the general situation and in turn told them about the state of the Navy of the Republic. Under the influence of events on the front, Madrid has clearly become more serious. On the streets and in the cafe there were fewer strollers, more armed fighters of the people's militia. They talked a lot about the wonderful activities of the 5th Regiment led by the Communists, which successfully formed military units and detachments.

As expected, I. Prieto was appointed Minister of Marine. He received me in his office at the Ministry of the Navy. Prieto sat at a huge table, and his small eyes bored his interlocutor from under his overhanging eyebrows. We talked with him about my stay in Cartagena and on the ships. He agreed with my suggestions and at the same time recommended taking part in the campaign of the republican fleet to the North, to the Bay of Biscay, about which I had not heard anything. Very pleased, I thanked Prieto. The Minister promised to give the necessary instructions to the fleet commander (Capitán de Fragata[21] Miguel Buiza was appointed to this post) and jokingly remarked that we'll open some champagne

[21] Commander rank in USN/RN parlance.

after the successful end of the campaign. With an inherent ironic smile, Prieto added: "If, of course, we meet."

This ended my stay in Madrid. Two more times I was in the capital, until the government moved to Valencia, but constantly worked in Cartagena.

Acquaintance with the Republican Fleet

Political situation in the Navy Command and Personnel

The picture of the state of the fleet was complex and contradictory. Previously limited previously by military regulations, political life on ships, in the base and in the city after the rebellion intensified with particular force.

The socialists, communists, anarchists and republicans intensified their activities. The socialists, having come to power in September 1936 and handing over the portfolio of Minister of the Navy to I. Prieto, did everything to extend their influence to the fleet. Anarchists even before the rebellion had a significant influence among the workers and the petty bourgeoisie of Catalonia, Valencia, as well as Murcia. Among the sailors there were many people from these areas. On separate ships, anarchists were even in the majority. At first there were few Communists on ships, only during the war did their number and influence increase.

The officers who remained loyal to the Republic for the most part supported the socialists. There were very few officers - anarchists and communists.

I would like to say a few words about the sailors of the merchant fleet. In their class composition they decisively differed from the officers of the Royal Navy.[22] This affected the very first days of the war. Most of the merchant ships of Spain remained on the side of the Republic. The captains of these ships acted heroically, considering the delivery of goods as important as the struggle at the front. The captains of such transports as "Santo Tome", "Magallanes", "Mar Cantabrico", who went to the Black Sea several times and who came to Cartagena under danger to their lives, deserve special mention. Or the captains of the small passenger ships "Ciudad de Cádiz", "Ciudad de Barcelona", etc., which transported children and goods from Cartagena to the USSR.

The officers who came to the warships from the merchant fleet were distinguished by their staunchness and devotion to the Republic. At first, they lacked military knowledge for commanding positions, but they were always excellent navigators and assistant commanders. So, the destroyer "Diez" was commanded by Jose Antonio Castro, a former captain of a merchant ship. On the destroyer "Ciscar", and then on the cruiser "Cervantes", the assistant commander was Rafael Mancheka, who also came from the merchant fleet; the navigator on the "Escaño" was the merchant sailor Eugenio Escudero, and on the "Almirante

[22] Royal Navy refers to the career Spanish Naval officers as the Navy held onto traditions from the Royal era stretching back centuries vs. the 1931 established Second Republic.

Miranda" was Alvarez Rubiera. These people could be safely put forward to command posts.

In the first two months while the Giral government was in power the political life of the navy was poorly controlled by the central government and local anarchists played an extremely important role. It was a period of very intense international struggle. The fleet was controlled by committees selected from representatives of various parties. All the time there were endless debates and discussions. Two newspapers were published in Cartagena - Cartagenero and El Liberal. The first reflected the views of the anarchists, and the second - the Republicans and socialists. The heated debate on the pages of these newspapers sometimes turned into open armed clashes in the streets. During military operations, as I personally observed during the fleet's campaign to the North, there was heated debate on the bridge of the flagship cruiser "Libertad" about how best to repel the attack of the rebel ships, if any does occur.

Anarchists brought a lot of trouble to the command and created additional difficulties. On the ships they were unruly sailors. The resounding phrase: "Win or die", which they liked to flaunt, was not supported by deeds. Once I happened to spend the whole day on the eve of going to sea on the battleship "Jaime I", where the anarchists had a special influence; meetings were held in at least three places on the ship, calling "to show what anarchists are," but nowhere was there any noticeable preparation of the ship for a campaign and a possible battle. Later, "Jaime I" sank in the harbor of Cartagena after the explosion that occurred in his magazine. Smoking in the magazine on "Jaime I" was normal, and the enemy could do anything on the ship, where order and discipline were rejected.

Desiring to retain the leading influence in the fleet, Prieto appointed his supporter Bruno Alonso as political commissar to the fleet commander. Bruno, in turn, sought to strengthen his position on individual ships, appointing socialist leaders as commissars.

As throughout the country, in Cartagena, on ships and in the base, there was an increase in the influence of the Communist Party. The party's clear and consistent policy in the fight against fascism attracted many to its side. If the navy ministry controlled the selection of leading cadres and did not allow communists to such posts then among the rank and file of supporters of the Communist Party there were more and more.

The commander of the republican fleet M. Buiza before the rebellion was the captain of an auxiliary vessel and did not belong to the caste of highborn nobles; Buiza enjoyed the trust of the personnel. Tall, thin, he did not like to talk much and when he decided to speak, he expressed his thoughts briefly and abruptly after which he again fell silent for a long time.

After the rebellion, Buiza began to command the cruiser Libertad. Then the Central Fleet Committee invited him to act as commander of the fleet. On September 2, by the order of the Minister of War, M. Buiza was appointed fleet commander. An officer loyal to the republican government, he, unfortunately did not have enough experience and knowledge for such a responsible post. Being a brave man he could direct his ship against a more powerful enemy, but it was difficult for him to competently manage the entire squadron in battle. Further events confirmed this.

The second figure in the fleet was Bruno Alonso. His role as a political leader in a complex political relationship was exceptional. Being called the General Political Commissar, he considered himself above the fleet commander and his signature was positioned in front of M. Buiza. His influence on personnel and non-naval affairs should not be minimized. After the liquidation of the committees and the introduction of the institution of political commissars in November 1936, B. Alonso, under the banner of restricting the influence of anarchists, pursued a narrow sectarian policy of struggle against the Communists, weakening the influence in the Navy of the Popular Front and strengthening the position of defeats from the old officer corps.

The chief of staff of the republican fleet was Captain de Navio Luis Junquera. A very sweet but weak man, he did not provide the necessary assistance to the commander. He could only timidly ask his subordinates to do something. He was not able to command.

The commander of the cruiser "M. Cervantes "was the Capitán de Fragata M. Ubieta. He was a fairly experienced officer and managed the ship well. Ubieta tried to maneuver between supporters of various parties, but did not break into a fight with the rebels. In November 1936, his ship, moored without cover in the open roadstead of Cartagena, was torpedoed by an unknown submarine[23] and for a long time went out of order. Prieto patronized Ubieta and promoted him to the most important posts: at one time he was even the commander of the fleet. At the end of the war, when the rebels captured the island of Menorca, he played a simply treacherous role.

The commander of the destroyer flotilla was the Capitán de Fragata Vicente Ramirez, an Andalusian, unusually talkative, sometimes even too noisy. On his bridge one could hear only his speech, sprinkled with strong naval expressions. Participation in the suppression of the rebellion put forward Ramirez, a very young officer, for responsible work. Mobile and energetic, he appeared on one or the other destroyer, and at meetings with the commander, his voice usually drowned out all other conversations. Ramirez had enough knowledge to manage the flotilla, but his disorganization prevented the necessary order. His name in the flotilla was Don Vicente, less often by his last name. The rank title was not in use.

The submarine flotilla was commanded by R. Verdia, who was an example of courage and devotion to the Republic. Commanding submarine S-5, Verdia was the only officer not embroiled in rebellion; he enjoyed the confidence of the team and was able to lead the personnel. Thanks to Verdia, all submarines remained on the side of the republican government, and he would undoubtedly

---

[23] Many years after the war it was revealed an Italian submarine Torricelli had done the deed.

have done a lot in the fight against the rebels if he had not died in the first months of the war in Malaga.

The commander of the base in Cartagena was, as already mentioned, Antonio Ruiz. A submariner by profession, he considered himself a Republican and a supporter of I. Prieto. A tall, handsome brunette, he did not like to do rough work and rather watched the events in the base entrusted to him than led them. His office and dining room served as a resting place for ship commanders with long, Spanish-style dinners and cheerful conversations. When Ruiz had the task of unloading transports with bombs and airplanes, he tried to entrust this "unpleasant" mission to someone else. And if huge stacks of explosives were concentrated on the berths near the base commander's building, Ruiz was nervous and did everything possible to quickly remove this dangerous cargo from Cartagena. In such cases, I thought: "There is no silver lining." I had to work side by side with Ruiz for many months. When I left Spain, he was appointed assistant naval minister.

The military governor of the base Cabrera[24] made the impression of a general who accidentally got stuck in a republican camp and waits for any chance to go to Franco. In Cartagena, he did not particularly interfere, and soon was generally recalled for work with Largo Caballero. I remember Cabrera well in connection with the next episode. Once, at a meeting with Prieto, not noticing my presence, he spoke out against Soviet volunteers; noticing me he blushed deeply and was embarrassed. Just a few days before in Cartagena, in his office Cabrera poured out in front of me compliments to the volunteers from the USSR!.... Subsequently, Cabrera was exposed as a traitor.

Now since these lines were written much has become clear. Another thing then in the first months of the civil war, some of the fleet officers who remained loyal to the republican government were guided at that time not so much by their sincere

---

[24] Toribio Martínez Cabrera (1874-1939) Participated in an anti-communist coup in 1939. Executed by the Franco regime.

desire to fight Franco but by the fear of being destroyed as his supporters. And only a few were conscious and sincere opponents of fascism and were ready to go against it to the end. While the scales were in the balance, the officers wondered how not to lose and did not know which position to take. Outwardly, everyone tried to emphasize their loyalty to the government of the Popular Front. As the affairs of the Republic worsened and the masses demanded decisive transformations within the country, officers increasingly leaned on the side of Franco. Class interests and caste prevailed. In 1937, hesitant officers under the influence of the general mood carried out operations to ensure communications, shelled the rebel occupied coast and more than once went out in search of an enemy squadron. In 1938 the situation changed. Among officers, pro-fascist sentiments and unwillingness to fight began to be observed more and more often. This was reflected in the direct disruption of some operations, and, in particular, in the disruption of an extremely important landing of troops in the Motril region. As the situation on the land fronts worsened and the confusion in government circles increased, as well as under the influence of international propaganda (England and France) in favor of Franco, some began to think about surrender. Such as M. Ubieta, threw off the mask and began to flirt with Franco. Even the commander of the fleet, M. Buiza, who did not raise any doubts at the beginning of the events (who, after a temporary resignation from his post in 1939, began to command the fleet again) at a meeting with Negrin spoke out for surrender, considering the situation hopeless.

There is every reason to believe that some of the naval leaders were associated with Colonel Casado, who led the conspiracy against the republican government. In such circumstances, a fascist rebellion in Cartagena on March 4 became possible.

The fleet, which had every opportunity to quickly suppress this rebellion, preferred to go to sea, hover out there for several days and then intern in Bizerte, among the French. If we recall that Buiza spoke out at a meeting of military and naval leaders in February 1939 for surrender, the conclusion suggests himself that he enforced this decision, taking advantage of the fact that the morale of the personnel was already broken.

One can imagine what a tragedy for sailors, non-commissioned officers and a small part of the officers who remained loyal to their people was the departure of the fleet to a foreign port when the struggle was still ongoing. As it is now known, a part of the naval officers-traitors turned to Franco with a letter in which they asked for mercy, assured themselves of their loyalty even in the days of the civil war, taking credit for the insufficient activity of the republican fleet. The enemy showed his face when there was no longer any need to disguise himself. However, do not let this insult the feeling of loyalty to the Republic of the bulk of the personnel, including some officers.

Of course, in the days of the rebellion a number of non-commissioned officers and sailors came forward, distinguished by their activity and political maturity. Of these, in the future, despite all the obstacles created by the naval minister and old officer corps, came capable gunners, miners[25], navigators, and even ship commanders came out. Unfortunately, these people often lacked knowledge and experience, which, of course, adversely affected the effectiveness of the republican fleet. Here we must look for one of the important reasons for some of his mistakes and failures.

However, before proceeding to a description of the combat activity of the republican fleet, it is useful to briefly talk about its composition before the rebellion.

Ships and bases of the Spanish fleet

In total, Spain had before the rebellion two old battleships, seven cruisers of various types and 17 destroyers. In addition, there were several small torpedoboats with displacement of about 180 tons that were of no value.

---

[25] Minelaying specialists. Soviet Navy offered three specializations for deck officers-Gunnery, Mines, Navigation.

Armed units also included five gunboats and about 20 small coast guard ships.

The Spanish submarine fleet consisted of six Type C submarines (newer) and six Type B submarines.

As can be seen from the above data, Spain had a significant fleet. At the beginning of the rebellion it was in varying degrees of combat readiness: the construction of two Canarias-class cruisers was being completed, some of the ships were at the docks or under repair.

The difficult economic situation of the country affected the timing of the ship construction, repairs were not done on time and the material part was not always kept in satisfactory condition.

The largest naval bases in Spain were El Ferrol and Cadiz on the west coast and Cartagena on the east coast of the Iberian Peninsula. The most modern base was El Ferrol. This base had the largest docks and shipyards.

An important base was also Cadiz, where for a long time the ships of the Spanish fleet were based and where there were many educational institutions and training grounds. Cadiz was a kind of cultural center and relic for the Spanish fleet, reminding sailors of the former sea power of Spain. The strategic importance of this base is evidenced by its close proximity to the Strait of Gibraltar.

Finally, Cartagena was a major naval base in the Mediterranean. Cartagena shipyards were designed for the construction of destroyers and submarines. The dock was lengthened when it was required to put the Cervantes cruiser there for several months.

In addition to the main bases, the Spanish fleet had many small ones: in the north - Bilbao, in the west - Vigo, in the Strait of Gibraltar - Ceuta and Algeciras, in the Mediterranean Sea - Mahon and Malaga. In addition, Spain had a large number of trading ports along the coast and on the islands; in case of need, they could also serve as a mooring and base for warships.

The personnel of the Spanish fleet and its training

The Spanish fleet and army were created especially in recent decades, not so much to fight an external enemy, but to fight the internal enemy. This explains the careful selection of servicemen of all categories, in particular, for service on the ships of the fleet.

Privileged officers can be divided into two main groups.

The first group is the commanding personnel of the ships, staff officers and base commands. These were the combat officers - the backbone of the monarchy. Therefore, representatives of aristocratic families and court nobles were especially carefully selected in this group. This situation has developed in the Navy over hundreds of years. The 1931 revolution did not make significant changes: the same admirals and officers remained in command of the ships. Nothing has changed by the time of the rebellion in 1936.

The second group included officers who held technical positions on ships, in arsenals and bases. This group mainly consisted of middle-ranking officers. Reluctance of aristocrats to engage in "rough" work forced representatives of the middle and petty bourgeoisie to assume the positions of engineers, designers, and mechanics. The second group for the most part supported the republican government, often openly expressing dissatisfaction with the aristocratic group in the ministry and navy.

Both groups made up the officer corps of the fleet, which differed sharply in their legal and material situation from petty officers and ordinary personnel.

Petty officers, i.e. non-commissioned officers who were selected from the ranks and underwent special training and education, consisted mainly of representatives of the petty bourgeoisie. However, among them were people from the ranks of the proletariat, especially where specialized knowledge was the main requirement. A large percentage of re-enlisted and career personnel were characteristic of this group. In submarines, the number of re-enlisted reached 65–70%. However, the officers did not succeed in "taming" the elders. Most of them were Republicans, Socialists or Anarchists. Sailors and petty officers of military service came mainly from industrial areas, workshops and factories and were representatives of the proletariat in the fleet. During the war, many foremen and sailors became supporters of the Communists.

Relations between officers on the one hand and petty officers and sailors, on the other were intensely hostile. The officers considered themselves entitled in assigning the main work to petty officers and sailors. Professional training for the officers was deficient. In order of seniority they received high posts and ranks but when commanding ships did not know how to leave harbors by themselves and, as a rule used pilots and tugboats. This, of course did not help strengthen their authority among the crew.

Remaining neutral in the war of 1914–1918 and without combat experience monarchist Spain was very slowly rebuilding its fleet in accordance with the experience and technical knowledge of the First World War. The Spanish fleet mainly engaged in campaigns in its waters, paying much attention to the ceremonial aspects and not to real combat training.

The combat training of individual ship formations was not coordinated. Submarines were built only in order to keep up with fashion, and they were not dealt with much. In the exercises, they did not interact with surface ships, their material part was in a state of neglect. The small hydro aviation was also not organically connected with the fleet.

All of the above explains the level of operational and tactical training of officers of the Spanish fleet and naval formations before the rebellion.

The requirements for the training of the rank and file were relatively low. A large percentage of long service personnel provided for the navigation of ships and conducted simple combat training.

Beginning of the rebellion

Turning to a description of the actions of the fleet during the period of rebellion, one cannot but describe the political situation in Spain. The 1931 revolution had a great influence on the personnel of the fleet. Supporters of various parties began to express their views more openly. While almost all the officers remained monarchist under the Republic, the entire rank and file went over to the side of the Republic; this further emphasizing the irreconcilable contradictions between the officers and sailors of the Spanish fleet. In February 1936, when the Popular Front government came to power, officers began to express their hostility to the Republic even more openly. In an effort to protect the rank and file of the Spanish fleet from the influence of revolutionary elements on the shore, from "politics", the command increasingly loaded the sailors with various exercises.

The Republican government made some organizational arrangements for the fleet: several admirals and officers, known to the whole fleet for their rude attitude towards sailors, were dismissed; Formal access to youth from the middle and petty bourgeoisie was formally expanded in naval schools; the position of the rank and file on ships both in material and legal terms (in the form of terms of service, shore leave, etc.) was somewhat improved.

However, these changes were too insignificant to create a truly new situation in the Navy; moreover, the commanders of the courts sabotaged them in every way. When at the end of 1933 right-wing bourgeois parties came to power, admirals and officers on ships launched an attack on the gains of rank and file that were achieved under the republican-socialist government in the first two years of the revolution. The February election of the Cortes in 1936, brought victory to the Popular Front[26], clearly showed that

---

[26] Authors Note. The electoral bloc of the Popular Front parties included: the bourgeois left republican party, the socialist, communist, etc. The largest group of

the revolution in Spain did not die, that there were all conditions for its further development. Desperate for the possibility of strangling the revolution along the paths of constitutional legality, the reaction led by generals and admirals relied on an armed rebellion against the Republic.

The rebels were assisted by fascist Germany and Italy, with whom the reactionary Spanish military had long established contact. Special headquarters were even created at the headquarters of Germany and Italy, which were engaged in the preparation and planning of the rebellion in Spain. However, the left-bourgeois republican government did not take effective steps to suppress the brewing conspiracy.

General Franco, convicted of preparing a rebellion, the day after the victory of the Popular Front was transferred as governor-general to the Canary Islands, and his accomplices – generals and officers – were dismissed with full salaries and could continue to prepare a conspiracy against the Republic score. On the territory of Portugal, the rebels created warehouses for the necessary stockpiles of weapons. The ships of Germany - a battleship, several cruisers and destroyers - were in the ports of Spain all the time in the summer of 1936, ready to provide the necessary assistance to the rebels.

Most of the officers were gradually involved in preparations for the rebellion, and under various pretexts, "processing" of the personnel of the garrisons of the army, aviation and ships of the Spanish fleet was carried out.

The activities of fleet officers a few weeks before the rebellion were largely subordinate to preparations for the uprising.

At specially arranged banquets, army and navy officers openly advocated "saving the homeland," and in the Canary Islands, General Franco defiantly refused to make a toast "For the Republic."

---

deputies in the Cortes (parliament) elected on February 16, 1930 belonged to the left republican party, which formed the one-party government M. Azana.

A few days before the rebellion, the commander of the cruiser "Republica", which was undergoing major repairs in Cadiz, informed the Minister of the impending uprising. He did not dare to report this through his officers, but took advantage of the premises and communications of the civil governor of Cadiz. The received radiogram was discussed secretly in the ministry, without the participation of the naval headquarters, because it was difficult to determine who to rely on. However, even now nothing serious has been done to suppress it.

On the night of July 17-18, 1936 a rebellion broke out in all the garrisons of Spain. True, the calculation of a quick and simultaneous coup in the whole country did not materialize. Despite the unpreparedness of the government, the anger and the will to fight of the broad masses were so great that spontaneously arising working groups of the people's militia crushed the rebellion both in the capital and in most industrial areas. In addition to Morocco, only in a few areas of the north and south did the rebels manage to seize power. But, divided into two parts, they were in an emergency.

What happened at that time on the ships of the fleet?

Even before the rebellion, the crew drew attention to the activity of the officers, to the unusual meetings of officers in cabins and bases. Radiograms passing through radio operators became the property of the crews and revealed the true state of affairs. The surveillance established by the crew (shore leave and other similar measures) aroused even more suspicion among the sailors against their officers.

In recent days, ships have openly talked about the impending uprising and the need to take appropriate measures against the rebels. Attempts by the officers to mislead the rank and file with various deceitful reports of unrest in the country have failed. On the ships there were many active members of various parties of the Popular Front, who urged sailors to be vigilant. And when the signal for the uprising was given on July 18, the rebels managed to deceive and lead the rank and file of several land garrisons, while the command of the ships remained loyal to the Republic.

All ships that were at sea or could leave the bases remained loyal to the Republic. Suppressing the resistance of the rebel officers, the ship crews notified the government of their readiness to defend the Republic. On separate ships at sea, armed clashes of officers with the crew occurred, ending, as a rule, with the liquidation of some officers. Thus, the reactionary rebellion in the fleet failed. Of the ships on alert, only the destroyer "Velasco" sided with Franco.

The fate of the ships located in El Ferrol was tragically formed. This concerned primarily two new unfinished cruisers of the Canarias type, the battleship "Espana" (in dry dock) and the cruiser "Cervera" (undergoing maintenance).

The rank and file of these ships was on the side of the government and in the early days of the rebellion stubbornly resisted the rebels. Unable to withdraw the ships from the base, the sailors first seized the territory of the base, the arsenal and the workshops. Bloody battles took place on the streets and in the harbor, but the fate of the base of El Ferrol was decided by the land garrison under the command of fascist officers. The base was captured by the troops of the garrison and the gendarmerie, and with it the ships that were there fell into the hands of the rebels.

As a result, the battleship "Jaime I", light cruisers "Libertad" and "Cervantes", the old light cruiser "Mendes Nunez", destroyers "Lepanto", "Diez", "Sanchez", "Ferrandez", "Churruca", "Galliano", "Valdes", "Antequera", "Gravina", "Escano", "Ciscar", "Jorge Juan", "Uloa", "Alcedo", "Lazaga", and all submarines remained on the side of the government.

In the hands of the rebels were: the battleship "Espana", the unfinished cruisers "Canarias" and "Baleares", the light cruiser "Cervera", the old light cruiser "Republica", the destroyer "Velasco" and all gunboats, with the exception of one.

Later, Italy "sold" four destroyers to the rebels: "Ceuta", "Teruel", "Melilla" and "Huesca", while German and Italian submarines often provided Franco with "friendly services".

Some details of what happened on the ships at the time of the rebellion are interesting. The assurances of the naval admirals and officers that they will lead the crew of the ships gave the leaders of the rebellion a reason to entrust the fleet with an important task: to ensure the transportation of troops from the African coast to the peninsula.

At the headquarters of the naval minister, all the leadership was involved in the rebellion and was ready to give the necessary orders to the ships to support the rebels. However, the circumstances were such that several signalmen lieutenants turned out to be loyal to the republican government and, instead of radiograms ordering them to join the rebellion, sent messages to ships at sea that Franco had rebelled and that the rebels intended to immediately transfer the Foreign Legion[27] to the peninsula. These messages were received primarily on the cruisers "Libertad" and "Cervantes". The radio operators on the cruiser "Libertad", having received such a message, did not transmit it to the command of the ship to which it was addressed, but to their political leaders from the crew, who called on the personnel not to obey the rebels.

In the early days of the rebellion, the radio cabin of the cruiser "Libertad" turned into a kind of command post of the fleet. While there was a struggle with opposing officers on the bridge and in the cabins, calls came out from the wheelhouse to all the ships to remain faithful to the legitimate republican government and not to believe the officers who were trying to deceive the team, claiming that the power had passed into Franco's hands.

The flag of the squadron commander was hoisted on the cruiser "Cervantes". The personnel of the cruiser, having received a message from the cruiser "Libertad", first of all blocked, so to

---

[27] Reference is to the Spanish Foreign Legion.

speak, the entire squadron's communication and on the way to Gibraltar sorted out the situation. Almost all the officers were arrested, and the crew decided to go to Tangier instead of Algeciras. Meanwhile, the situation in Madrid was in favor of the republican government, and the radio reported this. Eyewitnesses told how the commander, Admiral M. Mier, after a failed attempt to draw his ships into rebellion, walked the bridge in impotent rage. He allegedly hesitated at first, but when the question was directly raised before him whether he was ready to command the squadron on the side of the republican government, he refused and decided to join his like-minded rebellion. After that he was arrested.

After receiving a radiogram from the cruiser "Libertad", officers on the battleship "Jaime I" stubbornly resisted the Republicans, where it came to machine-gun fire from the bridge and severe reprisals against the officers.

And here is another episode told by an active participant in the events in the destroyer "Lepanto". Immediately after receiving news of the rebellion, the republican government sent a flotilla of destroyers consisting of "Lepanto", "Sanchez" and "Valdes" to Melilla (Africa) with the task of preventing the rebel troops from being transported to the peninsula. However, of the three destroyer commanders, two were rebels, and only the "Lepanto" commander, Valentino Fuentes, remained loyal to the republican government. "My officers rebelled," don Valentino, as his name was on the flotilla, addressed the team and offered to arrest the rebels. His order was executed.

The situation on "Sanchez" and "Valdes" was different. The lively negotiations between the commanders of these ships and the suspicious connection with the shore forced the crew to take decisive action against their officers.

An active rebel - the commander of the "Sanchez" - Bastareche jumped aboard the destroyer "Valdes" allegedly with a telegram in his hands and tried to call for rebellion; he said that there was an uprising in the country and Franco appealed to the patriots with an appeal to save his homeland.

Bastareche named the main internal enemies of his "fatherland" - the Masons and the Communists, against whom Franco primarily fights.

This, however, did not help.

Machinist Hevara interrupted him with a remark: "We want to know to whom you obey and from whom you received the order." The commander was at first confused, but then replied: "I obey the order of General Franco - the savior of the motherland." “We cannot obey any order, except for the order of the government established by law,” the machinist answered.

“We must not obey the order of the government,” insisted the commander of “Sanchez”, “by this order we must open fire against our brothers in Melilla. Which of us will raise a hand for this? ”But the strong voice of the machinist Rocha interrupted him: “In Asturias in 1934 there were also our brothers, but you fired at them mercilessly.”

“It is useless talking to them!” - exclaimed the rebel commander unable to bear it.

This decided the fate of the destroyers.

“Long live the Republic! The crew yelled. “We demand a return to Spain.” We will not take part in the rebellion! "

With great difficulties and with the help of a tugboat the destroyers left the rebellious port and returned to Cartagena. However, the ships approached this port with great caution because they did not know in whose hands it was, lingering at Cape Palos for a long time as they found out the situation. Only the next day, after running out of fuel and food, they entered the harbor. Cartagena remained loyal to the republican government. The personnel of the base and the workers of the city welcomed the incoming vessels.

So, one by one, the ships receiving information from Madrid and from the cruiser "Libertad" dealt with the rebel officers and joined the republican fleet. Cartagena somehow spontaneously became its concentration center and then the main base for the entire time of the war.

Of the military officers in the Navy, only very few remained loyal to the Republic. Here is some interesting evidence to prove this[28]:

Rank	Total number	Remained loyal to the Republic
Admiral	19	2
Captain I Rank	31	2
Captain II rank	65	7
Captain III rank[29]	128	13

Republican fleet in the struggle for Gibraltar

Having entrenched at the beginning of the rebellion in Africa and in the southern part of Andalusia, the rebels simultaneously seized all the ports and bases in the Gibraltar region. Both banks of the strait were in their hands. Cadiz and Algeciras on the peninsula, while Ceuta and Melilla on the African coast were Franco strongholds for control of the strait.

The Republican fleet was not able to prevent these actions by the rebels. Ships loyal to the Republic had first of all deal with the suppression of the rebels on board. Committees, as we know, took control of the ships and their protection. Hidden supporters of the rebels were on all ships and they needed to be exposed. It was necessary to understand the situation, to determine where to go for bases and replenish fuel, and obtain appropriate instructions.

---

[28] Author's Note. Benavides. La Escuadra la Mandan los Cabos. Mexico, 1949.

[29] Soviet naval ranks indicate, I, II, III respectively correspond to Captain, Commander and Lt-Commander in USN/RN parlance.

One cannot ignore the fact that the rebels had been preparing for the offensive for a long time, developed detailed plans, and all the leaders knew what they should do upon the signal of the rebellion. The new fleet leaders – committees and selected ship commanders – were new to the situation in the country that developed after the start of the rebellion. Radio data was inconsistent. Only in the bases it was possible to obtain reliable information about the events taking place in the country.

So, the cruisers "Libertad" and "Cervantes" were forced to call at the international port of Tangier in order to clarify which bases and ports are in the hands of the republican government. The battleship "Jaime I", which was on its way to Gibraltar, almost went to Cadiz, already in the hands of the rebels, and only information from the cruiser "Libertad" allowed the crew to correctly navigate and join the republican fleet.

In fact, the fleet was left to its own devices. The Government of the Republic, taken aback by the events and unprepared to suppress the rebellion, was unable to control events in the early days. Moreover, it could not solve such complex issues as the capture of one of the ports in the strait, for example, Algeciras, without which it was impossible to achieve dominance in the strait, but this required the combined actions of the fleet and ground forces. Meanwhile, the government had almost no regular troops left: military operations were planned and conducted by volunteer militia units formed by parties and trade unions, which were often guided by purely local considerations. For example, Catalan organizations, with the support of the fleet, landed troops in the direction completely opposite to that which was vital for the Republic. This landing on the island of Mallorca - after a month of fruitless effort - was undertaken in late August. Then it was still possible to organize a landing in Gibraltar, relying on the superiority of the forces of the republican fleet and the increased strength of government troops. But this time too, local considerations appeared in the government itself; guided by them, Minister of the Navy I. Prieto, a deputy from Bilbao, sent the entire fleet from the Mediterranean Sea to the North, to the

Bay of Biscay. Unfortunately, even the Naval Headquarters did not possess authoritative and competent specialists who could competently manage the naval operations of the fleet.

When the fleet returned to the Mediterranean Sea in mid-October, the balance of forces on the sea had already changed, and almost the entire African army of Franco was transferred to the peninsula.

The closest port to the strait, which remained in the hands of the Republicans, was Malaga, where the republican ships – the battleship, cruisers, seven destroyers and several submarines – concentrated in late July. The task of the republican fleet in the current situation was to prevent the transportation of the remaining troops from Morocco to the peninsula.

Having established the constant monitoring by destroyers and submarines over the strait, on July 23, the large ships launched their first combat sortie and bombarded Ceuta. The cruisers "Libertad" and "Cervantes" and the battleship "Jaime I" took part in this operation. The enemy's coastal batteries and port were shelled.

The rebels responded with coastal battery fire, and then attacked republican ships from the air.

On July 25, the same combination of republican ships fired at the port of Melilla, where, according to reports, troops and transports were concentrated to transport parts of the Foreign Legion to the peninsula. The destroyers and submarines strengthened the protection of the strait and did not allow any movement of rebel transports between Morocco and the ports on the peninsula. Systematic monitoring was carried out over the ports of Melilla, Ceuta and Larache, where destroyers and submarines were constantly on duty.

The rebels were not able to oppose the operations of the republican fleet. At that time they still did not have large ships operating, and air raids on republican ships were not very successful, but soon German and Italian military ships came to the aid of Franco. The German battleship "Deutschland" already during the first bombardment of Ceuta by its maneuvering interfered with the actions of republican ships. Then in this area there were already constantly German and Italian cruisers and destroyers, which limited the fighting activities of the Republicans.

Until the end of July, republican ships successfully blocked the strait, and during this time the rebels managed to transfer no more than 2000 men to the peninsula, and even then mainly by air.

On August 3, the battleship Jaime I, the cruiser “Libertad”, and the destroyer “Valdes” again fired at Ceuta. Large fires were noted at the port, but on the night of August 4-5, the rebels sailed out the first transports from Ceuta to Algeciras, covering them with their gunboats. The destroyers “Lepanto” and “Galliano”, who were on patrol, could not cope with their task and missed the transports.

On August 5, a large unit of the republican fleet consisting of a battleship, two cruisers and several destroyers entered Algeciras Bay and fired upon the port; the gunboat “Dato” was sunk, several transports were damaged, but the troops transported the day before were apparently already outside the port. In addition to Algeciras, republican ships fired at Ceuta, Tarifa, Cadiz and La Linea, where large fires occurred. The experience of war at sea shows that it is impossible to achieve decisive success with shelling alone. Only a landing and the seizure of one of the external ports or base could return the Republicans control of the Straits of Gibraltar. This could only be done with careful planning of all operations against the rebels. We have discussed this issue more than once in the light of missed opportunities, but we have always come to the conclusion that it is easier to reason retroactively when all the cards are revealed. “Yes, if not for the intervention, we would have defeated the fascists even without it,” Pedro Prado once told me and was to a certain extent right. But what was, was.

Rebel aircraft concentrated in the area of the strait and intensively bombed ships and the battleship "Jaime I" was damaged. The continued presence of the republican fleet in the strait became difficult.

The navigation of picket destroyers also became dangerous and the blockade of the strait began to weaken. In August the rebels transported more than 7000 men by sea, and in September more than 10000 troops and a significant amount of cargo.

The ships of the republican fleet, constantly at sea, required repair. In September, they concentrated in Cartagena and the battleship "Jaime I" began to be repairs at Almeria.

Thus ended the first month of the Republican fleet's struggle for Gibraltar. The crews of the ships, despite the absence of officers, showed that they can act and are ready to fight the enemy to the end. The transportation of troops so much needed by Franco on the peninsula was completely interrupted for two weeks, and this greatly helped the Republic.

Until the end of September, the Republican fleet continued to carry out systematic picket service in the strait and controlled Gibraltar. Unfortunately, these weeks were not used on the Central Front. Militia units, poorly armed and without experienced commanders, could bravely go into battle, but were not able to keep the enemy and, moreover, go on the offensive.

Republican Navy Northern Campaign

In the first months of the war, the operation associated with the dispatch of the entire fleet in September 1936 from the Mediterranean Sea to the Bay of Biscay in the North of Spain and its return to Cartagena deserve special attention.

What circumstances compelled the republican government to undertake such a risky operation, during which fleet ships had to break through twice through the narrow Straits of Gibraltar, by then already captured and fortified by rebels and twice to pass along the coast occupied by the enemy?

The decisive moment was the situation that prevailed then on the land fronts. The first weeks of the rebellion were critical for Franco: the rebellion was suppressed in a significant part of the garrisons, most of the aviation and navy were on the side of the Republic; the areas occupied by the rebels in the North and the South were separated by the strip of republican territory lying between them.

In September, the northern and southern areas occupied by the rebels were united with the help of Germany and Italy. About 50,000 reliable troops were transferred Franco from Morocco to Spain.

On the northern front, the rebels captured the cities of San Sebastian and Irun, thereby cutting off Asturias and the Basque country from France, communications with which were very important for supplying the population of the North with food and ammunition. In response to the insistent requests of the Basques and Asturians for help, the government sent most of the Navy to the Bay of Biscay. This was one of those poorly thought-out decisions, which, as indicated above, stemmed from the inexperience and lack of competence of the naval leadership of the republic. Further events fully confirm this assessment of the northern campaign.

In fact, the help that the fleet could and did render to the northern front of the Republicans was reduced to shelling the coast occupied by the rebels, delivering 2–3,000 rifles and temporary morale assistance to the population. This help was not able to prevent the loss of the North in 1937. Meanwhile, the northern campaign, associated with a great risk to the fleet, cost him the loss of one and the long-term failure of another destroyer. However, it was much more important that the rebels had the opportunity to relocate their cruisers to the south and finally secure the area of Gibraltar.[30]

The strategic expediency of the northern campaign is extremely doubtful, the campaign itself as a sea operation was very successful and deserves to be remembered.

In mid-September, the fleet received an order to enter the North.

The squadron of the battleship “Jaime I”, the cruisers “Libertad” and “Cervantes”, 6 destroyers – “Valdes”, “Diez”, “Antequera”, “Miranda”, “Escagno” and “Lepanto”, having received an order and discussing it in ship committees, as was customary at the time, prepared for the operation and concentrated on September 20 in Malaga, the southernmost republican port of the Mediterranean Sea.

The Republican fleet was commanded by Miguel Buiza, and the commander of the destroyer flotilla was Vicente Ramirez.

The Central Fleet Committee, together with the commanders of the ships, discussed the sortie plan and especially examined in detail the time, formation and connection courses during the breakthrough through Gibraltar. Everyone was in a fighting mood.

We decided that the force would leave Malaga in the afternoon to approach the Straits of Gibraltar in darkness: the ships, traveling in line ahead formation, would first approach as close as possible to English Gibraltar, and then turn to the African coast to avoid shelling of Algeciras batteries and aviation raids from Cadiz.

The further plan was to go to sea as far as possible during the night and thereby complicate the search for a fleet by rebel

---

30 Author's Note. A significant role in the decision to send the fleet to the North was played by the fact that this area has long been a pillar of the Spanish socialists, and its needs were especially close to the heart of Indalecio Prieto, who served as naval minister. Prieto repeatedly went to the Cortes from the city of Bilbao. The political leadership of Asturias was in the hands of socialists - supporters of Prieto.

aircraft the next day, and then, rounding the Iberian Peninsula, approach the ports of Asturias and Basque Country during the day.

The most likely meeting place with the enemy – the sea off the base of El Ferrol - was scheduled to be passed in the afternoon. Meetings with the ships of the rebels were not sought, but not feared.

The anchorage in Malaga roadstead was calm, except for two small and unsuccessful raids by enemy aircraft (bombs fell not on the roadstead, but in the city, although they bombed from a height of not more than 800 meters).

During this halt the cruiser "Mendes Nunez", who was at the time of the revolt in Rio de Oro, entered and joined the republican fleet. The personnel of the cruiser firmly decided to break into the Mediterranean Sea and fight on the side of the republican government; previously all officers, supporters of Franco, were released from the ship. No one reproached the crew for such an act. They told me that many officers from the" Mendes Nunez" later fought on the "Canarias" against their cruiser.

A loud "Viva la Republica!" From all the ships was a reward to the crew of the cruiser "Mendes Nunez". His arrival lifted the spirit of the crews of the ships going on the operation. The 10 large ships lined up in the roadstead and the 12-inch guns of the battleship "Jaime I" gave a certain solidity to the entire squadron, the mood of the personnel naturally rose even more.

In the evening of September 21, the large ships of the squadron formed into line ahead and went into operation under the protection of the destroyers.

At the van was the cruiser "Libertad", followed by the battleship "Jaime I", and the cruiser "Cervantes" closed the column. Destroyers guarded the large ships during the day, leading the way, and when approaching the strait, after dark, took place in the wake of the squadron. The ships were put on combat alert and readied for sea; the boilers were ready to give the fullest speed. The ships smoked heavily and therefore it was difficult to think about some sort of secrecy of movement.

Information about the installation of large-caliber batteries in Ceuta and Algeciras worried the commander. Soon a German cruiser appeared, who clearly followed the movement of the squadron, moving parallel to the republican squadron.

When the ships approached the narrowest part of the strait - between Ceuta and Algeciras - in darkness and among the searchlights of coastal batteries glided past them, silence fell on the bridge of the cruiser "Libertad". It seemed that a salvo should follow, when the searchlight beam stopped for one second on one of the ships, but the batteries were silent: the ships passed out of reach of the guns or went unnoticed.

M. Buiza walked nervously along the bridge, more concerned with other ships than the cruiser "Libertad".

Behind the stern of the squadron, brightly lit English Gibraltar remained. Moving away from Algeciras, the sailors felt relieved: Ceuta was already behind, and the batteries of Algeciras did not bother the commander.

The tense state gradually subsided. Ships continued west. The shores of Africa and the peninsula disappeared. Ahead was the expanse of the Atlantic Ocean. One can imagine how hard it was for Republican sailors to pass by their bases and especially Cadiz. After all, until recently they were welcome guests there. In Cadiz everyone had acquaintances and some had families. And now, just from there, one had to expect the appearance of enemy aircraft.

Cadiz is not just a naval base for all Spanish sailors. The development of the navy and the heyday of Spain in the 16th century are associated with this city. Columbus twice left Cadiz on his distant voyages across an unknown ocean. There arrived rich caravans with jewels from America. It is no accident that the first meridian passed through San Fernando near Cadiz (later England laid the first meridian through its Greenwich). Galley and sailing ships left Cadiz during the years of greatness of the Spanish fleet. From there came the famous "Invincible Armada" in 1588, which was later defeated off the coast of England, more likely by storms than by the enemy. But history already worked against feudal Spain and the last attempt to hinder the development of the industrial fleet of Britain in 1805 ended in defeat at Trafalgar. Napoleon never loved or appreciated the fleet which may be why Villeneuve who led the Franco-Spanish squadron was utterly defeated by the brave Nelson.

At the time described Cadiz was a relic of the Spanish fleet as well as a training center and an excellent base.

The next day, September 22, the squadron was already far from the coast and busy sea routes. Dead swell from a storm somewhere past rocked the ships, marching in service under the protection of destroyers. The battleship was at the center of the group, creating pride and admiration among the participants of the campaign. His active fighting in the early days of the rebellion and huge 12-inch guns created a well-deserved authority in the Navy. True, the majority of the personnel underestimated its obsolescence, inappropriateness to modern naval combat and the futility in the event of a meeting with high-speed cruisers. Fleet ships - cruisers and destroyers - would be much more efficient without a battleship in this campaign, but the moral factor prevailed. The timid offers to leave him in Cartagena were rejected most emphatically. Anarchists, of whom there were quite a few on "Jaime I", sharply opposed such intentions.

All day, September 22, the ships calmly went on the planned course. Airplanes did not appear and the enemy's ships at that time were not at all afraid.

Towards evening the squadron turned north and Cadiz was again on the starboard beam, now at a great distance from the squadron.

The closer the squadron approached the El Ferrol rebel base, the more serious the commander Miguel Buiza became and the more movement could be observed with guns and equipment. Non-commissioned officer F. Mira, who served as the senior gunner of the cruiser "Libertad", was almost constantly at his observation post.

Vigo passed - at that time a small rebel naval base and once-famous military port in which the entire Spanish fleet stood competing with the British. Russian ships often entered Vigo, heading from the Baltic to the Mediterranean Sea, and sometimes stood there for a long time doing combat training, eliminating problems and replenishing the necessary supplies. The population of this city got used to Russian sailors as their own and the Russians jokingly called it not Vigo, but Viguysk.

Now Vigo, like El Ferrol, was in the hands of the rebels. A cruel reprisal was inflicted on republican sailors here.

By evening El Ferrol passed. It was necessary to turn east, around the Punta de Estaca de Bares, the next day to proceed to Gijon and Santander.

Nightfall and dawn are always the most critical time for the commander of a formation at sea: you need to make a decision for the night, based on the situation, reposition ships and check combat readiness.

Night meetings with the enemy at sea take place with lightning speed and are fraught with serious consequences. Each minute of delay can decide the fate of the ship.

But when the night ends the dawn prepares new surprises and the commander must make appropriate decisions. In these evening and predawn hours the commander is usually on the bridge carefully studying the situation.

The normal execution of duties which we have already become accustomed over the previous quiet days was unexpectedly disturbed by the cry of a signalman: "Aviones!" (Airplanes).

Military alert aroused all the personnel. One by one, the ships were put on full alert. Anti-aircraft guns quickly rotated, looking for a target.

The reconnaissance airplane did not approach the ships, crossed their course to establish the direction of the squadron. Opening fire was pointless. Soon, the airplane disappeared behind the stern, having received the necessary data on the movement of the republican fleet.

It was hard to predict what would follow this intelligence. The approaching twilight now bothered the commander even more and he nervously walked along the bridge examining the ships' formation. The horizon was getting darker - the enemy could appear from any direction.

We passed the most dangerous area and most likely meeting place with rebels. The time has already passed for turning into the Cantabrian Sea, but the squadron went to the northeast until finally convinced that the airplane had flown off.

It got completely dark when the ships following in a closer formation turned east. Full combat readiness was maintained on all ships until the excitement caused by the appearance of an enemy aircraft finally subsided.

On the bridge of the cruiser "Libertad" which housed the Central Fleet Committee, the supreme fleet management body in those days, there was a special revival. Various assumptions were made about the events for the forthcoming night. Everyone had the right to speak and no one had the right to stop the speakers. Least of all spoke the commander M. Buiza. He only occasionally inserted comments, besides not insisting on their approval.

Everyone was worried about possible air raids because there were rumors that Franco received a large number of new Caproni aircraft from Italy. But since this was possible only during the next day, because at that time airplanes still rarely appeared above the ships at night, the passions that had gradually flared up began to cool.

The squadron was gradually moving away from the dangerous area and the main base of El Ferrol, where there were cruisers and the battleship of the rebels.

Here is the Cantabrico. The night passed quietly, waiting for the appearance of the mountains of Asturias. The first port where all ships were to go except for a few destroyers was Gijon. The squadron laid down on a course leading to the harbor.

In the days of the campaign I managed to get to know the commander, Central Fleet Committee members and individual non-commissioned officers more closely. As when I was on the "Antequera" almost a month ago I was convinced how much enthusiasm the personnel had. But involuntarily I had fears this enthusiasm might not have the desired effect if the fleet did not have a solid order, and its leadership had sufficient knowledge. After all, people shoot, not guns and those who are better prepared are more likely to hit the target.

"A great non-commissioned officer," Buiza once told me, pointing to Mira who was passing by. He served as the gunner. And to my question, does he know how to control fire, Miguel spread his hands: they say he never had to.

It struck me a lot. In my cruiser I would consider smoking an extraordinary event during the loading of ammunition, and then here I met a smoking marinero, on whose shoulder lay a shell. But my charter in a foreign monastery must be offered carefully[31], and while I was silent, what else could I do?!

From Gijon to Bilbao

All the ships of the squadron proceeding directly to Santander, except for the three destroyers with the "Lepanto" at the head, under the command of Valentino Fuentes, which could be barely housed in the small port of Gijon. Even on a sunny, clear September day, the city looked grim. It was somewhat reminiscent of Cartagena. There are no tall bright houses or wide boulevards, as in Valencia and Alicante. The gray buildings here are pretty monotonous. Obviously, it is impossible to preserve vibrant colors where there is a lot of coal dust.

The berths of the port and the adjacent streets are filled with people. In northern Spain, people are calmer about events. It was felt now. Everyone was clearly pleased and had a lively conversation but no one shouted or rushed to the ships as was the case in Malaga. In addition, the proximity of the front left its mark and with the advent of the ships rumors appeared about a possible bombing of the city.

Representatives of the local government arrived at “Libertad”: Belarmino Thomas, a socialist from the working miners, and with him commissar González Peña. The commander had not yet managed to leave the bridge and the conversation took place there. After greetings and congratulations on the successful completion of the campaign, the local authorities and members of the Central Fleet Committee turned to professional discussions. Among various topics, two caused particular excitement: the delivery of weapons to Asturias and the belief that in the coming days it would be possible to occupy Oviedo, the capital of Asturias, which was surrounded on three sides. “This must be done before the autumn rains,” said González Peña, who as the commissar led the siege of Oviedo. The city’s occupation date has already been set several times, but so far they have all broken down. The situation of the besieged city was really difficult, and the commander of the defense, Colonel Aranda, even informed General Franco that he would have to surrender if he did not receive reinforcements. The position of the Asturians was not alarming. They had enough weapons and ammunition, and they only complained about the lack of certain types of food

---

[31] A paraphrase of a Russian proverb meaning to tread lightly when offering differing opinions on a topic compared to what is customary.

that Asturias used to get from Castile and Aragon. However, the Republicans did not have enough military knowledge and organization to close the ring of blockade and force the rebels to capitulate who stubbornly resisted waiting for an improvement of the situation.

Upon learning that I was a "Companero Ruso", Peña came up to me, shook his hand firmly and spoke of the help of the Soviet Union to the Asturians. I didn't understand much then and only guessed that it was a question of granting asylum to the Asturian miners after the massacre in 1934. He invited me that day to come to Oviedo and see how a struggle was going on the very outskirts of the city, which should soon end in victory. Without delay we really went there accompanied by Belarmino Thomas and Peña. M. Buiza refused to leave the ship until the delivered weapons were unloaded and P. Prado and I decided to use the invitation. We were advised to always wear boots and raincoats: boots should be worn because it is dirty in the trenches, and raincoats for disguise. None of this was on the ship and we were promised to get everything at the commandant's office when we would change to cars adapted for driving on bad roads.

From Gijon to Oviedo is about 30 kilometers. On the way we got acquainted on the map where the front line is, where the Asturians are going to advance and how the only road connecting the Oviedo garrison with the territory of the rebels is being shelled. Soon we stopped near a small gray house where the commandant's office is located. We changed clothes, moved to other cars and moved on. To get to the outskirts of Lugones, from where Oviedo was best seen and where the command headquarters were located, we had to drive several hundred meters through the open area, which was under fire. The road first went along the gorge. In the valley, "military cunning" was used. Knowing there would be a lull after an increase in fire, the cars stopped and then at the order of the experienced, choosing the right moment, we rushed at full speed crossing the dangerous place. Naturally, no engine breakdown was envisaged at this time. When both cars were safe, shell explosions were observed almost where we had just passed. But they say there have never been victims.

From here, from Lugones, Oviedo was viewed perfectly. A tall cathedral and a number of buildings in the city center were visible. We walked closer to a steep cliff where an observation post was organized in a building with thick walls. Through binoculars and telescopes inserted into narrow slots we could see not only the streets and individual houses in Oviedo but also people who quickly ran alone from one hiding place to another. Shot through from all sides, the city has already adapted to the conditions of a long siege. The excess population had been evacuated, the walls and windows of the houses facing the front were closed and fortified, the city looked empty. Meanwhile, life continued in him. Several lines of trenches on either side clearly marked the front line. González Peña told us that the rebels here in the northern section do not advance - the terrain does not allow; Republicans are not active here either. But in the west, the fascists have repeatedly tried to capture the small town of Truvia - it has a famous arms factory and various workshops. The city has been evacuated just in case, so that the plant does not fall into the hands of the rebels.

We thanked them for their hospitality and began to gather back to Gijon. When leaving the building the following happened: Pedro Prado inadvertently stepped out into the open area, and while he was dragged by the sleeve, several bullets were already whistling around him. "Every inch is shot through here," they told us, "and you cannot go out into the open during the day." And at the same time, having traveled two to three hundred meters, we noticed children playing carelessly. They were under the cover of a mountain. Here, everyone is accustomed to the sounds of cannon fire and shell explosions. Residents chose to stay in their home and did not leave. Here we met soldiers who came to rest from damp trenches and ditches. Nearby, in the lowland, all houses were destroyed and the civilian population was evacuated.

At the end of September there was a lull in this sector of the front. Hot fights took place here a little later - on the 10th of

October. The Republicans, advancing, broke into the city and it seemed victory was already ensured. But the attack was stifled and the Republicans went on the defensive, later leaving their positions. The direct witnesses to this unsuccessful attack were M. Koltsov and R. Carmen. I had something to hear about this from Koltsov himself, when a few months later we reminisced with him, visiting V. A. Antonov-Ovseenko, about the northern provinces. Mikhail Efimovich was surprised to learn that I participated in the campaign of ships to the North, traveled from Gijon to Bilbao two weeks earlier than him and like him flew to Madrid over the entire territory occupied by Franco (about his stay in Gijon and Oviedo Koltsov interestingly relates in his "Spanish Diary"[32]).

Having returned from Oviedo I was impressed for a long time with what I saw there. I recalled this later when I received information about the stubborn battles taking place in the North. Separated by trenches and barbed wire ordinary Spaniards stood against each other. What drove them to internecine warfare? How to explain such stubbornness of Franco's supporters? The enthusiasm of the Republicans was understandable. Under pressure from the masses, the monarchy was overthrown. In 1934, the people opposed attempts of reactionaries to restore the old order. The Spanish people had a definite and clear goal; they were enthusiastic and determined to bring the revolution to the end. However, as revealed in the very first days of the war, enthusiasm alone was not enough, organization and discipline were still required, weapons and the ability to use it in battle were needed. This is exactly what the Republicans lacked. That is why Oviedo, surrounded on three sides, remained in the hands of the rebels.

A different picture was observed among the rebels. The broad masses remained on the side of the republic; they did not want to support the fascists. This is entirely confirmed by the fact that the crews of the ships that fell into Franco's hands did not enter into

---

[32] Koltsov's Diary has been commercially published in Russian and Spanish text.

battle with the Republicans, even with a clear superiority of forces. It took time to change the situation with the help of German instructors and deceitful propaganda. The low political and moral state of the rank and file fighting on the side of Franco had to be compensated by equipment and instructors from Germany and Italy. Without open intervention, the rebellion would be quickly suppressed.

In this struggle, almost all the generals and officers were on the side of the rebels. They had the knowledge and ability to fight. We must not forget about the mistakes made by the Republicans which helped Franco conduct propaganda against the government of the Popular Front. Anarchists played a negative role here; their inconsistency and thoughtlessness of decisions regarding the socialization of land and property or the collection of industrial enterprises without sufficient preparation for this frightened the masses. The rebels skillfully used this.

Even in September the north of Spain is full of rains, the nights are cool and the trench war here is much more unpleasant than in the south. The winding mountain roads, pressed to the sea, small towns and villages that took refuge in gorges with passes and, finally, a humid sea climate - all this reminded me of trips along the roads of our Caucasus, somewhere in the region of Sukhumi and Batumi, in the rainy months.

In the same way we returned back, saying goodbye to our front-line masters. “Hasta luego” (goodbye), we said to each other, but we no longer were to meet. The next day, I went to Santander and Bilbao with a member of the Central Fleet Committee of Khativa, where I was supposed to get acquainted with the situation in the North and talk with some representatives of local authorities.

We stayed at a small hotel in Gijon for the night. It was packed with refugees from Oviedo and the two of us were accommodated in a room on the ground floor. The mood of the people of Gijon fell noticeably towards evening as they waited for an air raid. Gijon had not yet experienced major raids and the squadron that came here brought both joy and anxiety. We decided to go to a small restaurant for dinner and taste the famous Asturian cider. We were served meat, fish and some bread. With food it became harder every day. The tall Asturian set three large mugs on the table. "Look, look," Khativa turned to me, nodding his head toward the minister, who, taking one mug in his left hand, raised his cider bottle above his head with his right and deftly put a thin stream of the drink into the mug. "Why is this?" I asked, perplexed. It turns out that the cider in the bottle is still a kind of semi-finished product, but after the jet hits the edge of the glass or mug from a great height, the cider becomes foamy and finally ready for drinking. The drink turned out to be quite strong, contrary to my expectations. Two drunk mugs made themselves felt and we forgot about a possible air raid. Fortunately, there was no raid.

Early in the morning, barely dawning dawn, we went outside to go to Santander. The city slept calmly after all the anxieties of the past day.

There were warships in the harbor of the port. They were preparing to conduct a series of operations to shell the flank of the enemy army east of Bilbao and ensure the movement of transports with food and fuel. When we drove past the harbor, "Reveille" signals sounded on the ships – it was 6 o'clock in the morning.

An old-fashioned Mercedes, ugly but spacious, headed for Santander. We had to drive about 200 kilometers. The road then rises in sharp turns then runs along a flat plateau along the sea. Deserted places in the highlands are replaced by picturesque corners with "pueblos" and small towns. Khativa did not speak French well but I still understood little in Spanish and having "talked" another hour, both, tired, became silent for a while. Then it all starts all over again. I still need to learn Spanish, and I, not embarrassed by my mistakes, ask my interlocutor one question after another. I do not always understand his answers and again try to find out something, pleased when I succeed.

Here is Santander. The first impression is that we came to another country. Common to the eye in Cartagena and Gijon, the mono with zipper and beret suddenly gave way to exquisite costumes and hats. The day turned out to be sunny and the streets were lively. Despite the approaching twilight, no one was afraid of the air raid, and the walk on the embankment continued. On a wide spit leading to the former palace of the king cars rush and walk pedestrians. There are three destroyers in the port. They are getting ready to leave for Bilbao. They attracted the attention of the public, but the ships here were treated unfriendly, as if they had violated the unwritten neutrality in the war that the city adhered to. "Maybe we should fire at Santander first and then go to Bilbao," the sailors joked, but the serious discontent which this rich, well-fed and, moreover, still shocked populace by their arrival, slipped through these jokes.

One of the destroyers is our Soviet comrade N.P. Anin, and I propose to go to him to agree on further joint actions.

We climb the ladder aboard the "Lepanto". Valentino Fuentes, already familiar to me, is a little dry, standing on the upper deck and examining the sea with binoculars. He made a path very characteristic of many officers of the republican fleet. After the rebellion, he remained on the side of the government. He behaved well in the first year of the war, picked up by the general wave of anti-fascist struggle. Although Fuentes did not possess the necessary qualities, Prieto appointed him to command the fleet in Cantabrico. This was already in the spring of 1937. When the rebels attacked in September - October 1937, Don Valentino was at a loss, began to hesitate, not knowing what to choose. But this is all in the future. In the meantime, he commanded a destroyer, was a fairly experienced commander and was respected by the team for his services in the early days of the rebellion. He is clearly pleased with his stay in Santander. This city revived in his memory the years of youth, when he, in the rank of teniente (lieutenant), took part more than once in the pleasure trips of the ships accompanying the royal retinue to the summer residence; while naval officers in embroidered uniforms were more engaged in balls than in serious combat training. He hoped for a victory over fascism and apparently, sincerely wanted it. However, he rather peculiarly imagined the Republic. It seemed to him that the war would end and almost the old order would be established in the country and he would still occupy a privileged position. He did not think about the reforms that, as a sign of the times, were already knocking on the door of Spain with its feudal remnants.

Like most officers, and not only officers, he was not prepared for major social changes. Later, seeing how far a revolution could go, Fuentes drew back from it.

"Ola, don Nicholas," he greeted me, taking me under his visor.

Don Valentino was always distinguished by his hospitality, he knew the intricacies of etiquette and it was no accident that he represented the fleet at various ceremonies, which he liked to talk about.

At dinner, I spoke with Kompanero Ruso Anin.

We decided to continue on the next day. The driver complained about the engine of his clunker, and we were tired after spending more than seven hours in the car.

In a luxury hotel we were reluctantly provided with two rooms and somehow looked askance at my friend in mono and with a gun behind his belt. I looked like a more peaceful person. The civilian gray suit, albeit with a stretch, but still met the requirements of a fashionable hotel.

The manners of this city were amazing at that time, and not only of this city. Just as the crew of the "Mendez Nunez" cruiser, stationed in Rio de Oro allowed the rebel officers to leave the ship and without any incident to go ashore (and they immediately went to Franco's camp), here in Santander, instead in order to restore order among overt monarchists, they were given the choice of either a Republic or a foreign passport. Strange, but true. These rich people associated with London received passports, exit visas to England and calmly moved on British passenger steamships, which, at their request, brought them to England or to Vigo, i.e., directly to the rebels. This went on for several more months. In September 1936, there were especially many such people gathered in Santander. Ever since the time of Alfonso XIII, they preferred to spend the summer at this resort and were caught up in events here. They did not hide their views, and local authorities looked at everything through their fingers.[33]

When we went down to the restaurant in the evening, we were struck by the rude tone of the footman and the undisguised hostility of the public. At first I did not understand that a scandal was brewing and sat calmly waiting for the camarero to serve the menu. But the menus were not served. It turns out that one of the visitors protested against the presence of a sailor (it was about Khativa) in mono, and even with a gun behind his belt. Only by tone and decisive gestures with a menacing pat on the holster did I realize that things had gone far. I tried, firstly, to find out what was the matter and secondly, I asked to avoid conflict in my presence. It was useless for me in the position of attaché. Visitors to the table next to us ostentatiously went out and the owner who came running persuaded Khativa to calm down, promising to serve everything immediately. The loud conversation on this stopped and we, accompanied by general disapproving remarks, without hurrying, finished dinner and went out. According to Khativa, "the Venezuelan consul demanded to take us out of the restaurant", but the owner did not dare, as Khativa threatened to call the sailors and "put things in order" in the restaurant. "If it weren't for you, I would have expelled all these fascists with my marineros," the sailor, a former petty officer, and now a member of the fleet committee, continued calmer, but still angry.

Angered by the methods in Santander, Khativa suggested walking along the promenade. By evening all life in the city had moved to cafes, restaurants, and entertainment venues. Music played and little reminded of the war. But just 300 kilometers from here, we only yesterday watched how the Asturian miners in the trenches,

---

33 An expression from Tolstoy's Anna Karenina. Meaning to knowingly ignoring something reprehensible.

knee-deep in the water, besieged Oviedo and the inhabitants were ready to give the last blanket to warm the fighters. But there were as many such contrasts in Spain in the first months of the war.

In the morning we moved on. The weather turned bad and drizzled lightly. The high Iberian mountains as they approached Bilbao loomed ever closer to our right. There is more greenery and more water. Somehow unexpectedly we drove up to the control post in Bilbao. We showed their "salvoconducto" (passes) and moved on.

It was already the capital of Basque Country. Some information about her aroused some interest. They said that the language spoken by the Basques is similar to Georgian. Why? No one could give an answer. Situated in the mountains, Basque Country has maintained its identity, despite invasions by Carthaginians, Romans and Arabs. It has long been striving for autonomy, and only now is the government's decision on this issue expected from day to day.[34]

The front runs east of Bilbao, sixty kilometers away. After the rebels captured Irun and San Sebastian, there was a lull. Only the Italian "Caproni" troubled the city, but there were no heavy bombings yet. The city reminded me of Leningrad. In Bilbao, unlike other major cities in Spain, there are no tall buildings. There are many bridges over the Nervion River, which perhaps made me think of Leningrad, and the weather is also cloudy, with clouds hanging over the roofs of the houses and a light rain falling for weeks. Black dominates: black mono, black berets, women dressed in black, in black scarves. This is not mourning, but ordinary Basque clothing. Ships have not arrived yet and we seek refuge in a hotel. Like Gijon there are many refugees here, but from Irun and San Sebastian, all the hotels are crowded and we are arranged in one small room with the condition that we do not stay long. The Basque Communist Party General Secretary, Juan Astigarravia, whom I wanted to meet left town. He is busy forming units and went to the front. There is no

---

[34] Author's Note. The decision was made on October 1, 1936

Governor Aguirre. We are received by one of his deputies, a low broad-shouldered basque and talk about the situation in the city. He talks animatedly about the help that ships can provide if the enemy goes on the offensive. The Basques themselves have so far thought only of the defense of Bilbao. They have a lot of people but very few weapons. English and French transports export ore but do not bring weapons or even food. The owners of the mines send their products and leave the capital in a London bank (I recall Santander). In the morning we learned that three destroyers arrived at night and are at the port. A huge harbor with docks and shipyards. Many transports are moored: the Spanish are afraid to go out, and the foreigners ship out only what is in their interests.

I learned almost nothing significant in Bilbao. A pity, because there was not a single Soviet comrade here. I entered into the notebook some data on the situation at the front. There was no complete impression of the huge city. I had a chance to learn more about it later and while I was in Cartagena, I only imagined with what stubbornness the Basques fought for their capital, gaining autonomy in early October 1936. They fought in a difficult international political environment. From Catholics to Communists, everyone united in the fight against fascism but the general situation of the North, cut off from the resources of the whole country, unable to get help (after the fleet left) by sea, was difficult, and the Republicans, defending each city, nevertheless retreated.

On a rainy day, we returned to Santander. There was an opportunity to fly to Madrid. Having firm instructions "to come to Madrid as soon as possible", I decided to fly over the territory of the rebels.

Over the territory of the rebels

"You should return to Madrid at the earliest opportunity," M. I. Rosenberg told me when I was preparing to go to the North. There was still no talk of the return of the fleet to the

Mediterranean Sea. Therefore in Gijon as well as in Santander and Bilbao, I was looking for an opportunity to return to the capital. The shortest way is by air. From time to time planes flew from Bilbao to Barcelona via France and sometimes directly but this was rare. Occasionally, taking advantage of altitude and cloudy weather, a passenger Douglas risked heading from Santander to Madrid through the rebel provinces. I got lucky. Khativa reported that yesterday an airplane flew in from Madrid, which returned a day later. He was also about to fly and asked for my opinion. "What could be better than flying straight and with a good friend," I joked and, of course, agreed. In the evening, at the hotel, he specified the departure time - 8 o'clock in the morning. I put in order some notes necessary for the report to Gorev, and Khativa and I went to the city.

I didn't walk. Thoughts were in the way. "It will be cloudy and rain tomorrow, which is good," said Khativa, looking at the sky. We have already talked about this. In the case of clouds, everything seemed simple. The plane rises to a great height and under cover of clouds, quietly flies over the territory of the rebels. Unfortunately in September rain is rare in this area but we may be lucky. Vamos a ver - as the Spaniards say. Wait and see.

The next day, as soon as we had time to wake up some local government official arrived at the hotel and reported that the plane was flying on time and the car was ready for us. On the way to the airfield we make sure that the clouds hang low over the city and the sea unfriendly rolls its waves ashore and the gray horizon merges with the sea somewhere very close to us. "It seems we will fly under the cover of clouds," I say, encouraging myself and guessing the thoughts of a comrade.

"Segurissimo" (for sure), he replies, and with a gesture of his hand shows how we take off, quickly gain altitude, and then, flying "dos horas", that is, two hours, we land in Madrid.

The airfield is covered with tall green grass. There is no concrete path and the take-off field rests directly on the sea. The only Douglas stands at the very airfield hanger, but near the plane there is no activity as would be usual before departure or arrival. We arrived early. The crew has breakfast and the rest of the passengers have not arrived yet.

We were invited to drink coffee in the small room at the head of the airfield. Here we met with crew members. We get to know each other and bound by a common fate for the next two to three hours we behave like old acquaintances. I only remember the last name Garcia, Juan Garcia. After breakfast, the crew noisily rises and goes to the plane. A tall man in a leather jacket and uniform cap, a pilot, I guess, exchanges a few phrases with Khativa and sits down at the table with him.

From what I've heard I conclude that we are talking about weather, altitude and the danger that we may encounter. I see how, on an unfolded small map, the pilot shows with a pencil the path of the plane, which will gain height above the sea, then cross the mountain range separating the territory of the Republicans from the rebels and head for Madrid, bypassing the large cities occupied by Franco. From Santander to Madrid, about 400 kilometers and, therefore, in two or three hours we will be there. The pilot supposed to get around Burgos, lying just in our way. In case of meeting with enemy fighters, several machine guns were installed on the plane. The pilot warned that at an altitude of 5000 meters it would be cold and we should be dressed warmer.

The clouds rose a little but still covered the whole sky - this calmed us. At the last moment, the other three passengers arrived. Everyone's faces are serious. A flight carries a certain risk.

At the airport not a single plane and no launch team. The "Douglas" runs up, rolls for a long time on the green grass and almost off the coast lifts off the ground. It can be seen how small waves beat against the shore.

Soon the plane entered the clouds; now only by altimeter it was possible to find out that the pilot was rapidly gaining altitude -

400, 500, 700 meters. U-turn and the plane lies heading south. Small drops appear on the glass. I think: is it rain or just moisture from clouds. It is certainly safer to fly in the clouds over Franco's territory but it is depressing not knowing where you are and relying only on instruments and the navigator's calculations. Suddenly, a blinding sun fell over the cabin, and below us, like cotton wool, spread white clouds. Cantabrian mountains appeared ahead, into which the plane seemed to crash if it did not have time to gain altitude. In fact, they are still far away, and we, without changing course are flying forward. It is getting cool, and I enjoy using the leather coat that was offered to me. On the shelf is my small suitcase and it contains all the notes. Gloomy thoughts come to mind: what if we make an emergency landing on the territory of Franco? Should I destroy unneeded papers? What will I say if I, the only Russian is asked how I ended up on this plane? I do not find an answer and try to drive away these gloomy thoughts. To our chagrin, the weather forecast did not materialize. With the mountains passed over the clouds cover was cut off and below us is the territory of the rebels. Instead of greenery which the entire coastal North of Spain is so rich in, stretched plains scorched by the sun or gloomy mountains, as in Cartagena. Motors are strenuously humming, making maximum speed. Everyone looks down, but it's hard to see anything from that height. We try to talk and shout out the noise of the motors, but this is difficult to do. Behind the tense jokes, one feels a general concern. Khativa introduces me to one of the passengers. This is a Basque named Echevarria. He occupies a high post in Basque Country and flies to get real help from the central government with weapons and food.

"We will fight for Bilbao to the last man," he said convincingly with ardor, "but we have few weapons and bad food." He is indignant that France does not support the Republic while German and Italian planes openly participated in the attack on Irun and San Sebastian. "These cities were not occupied by the Spanish rebels but by the interventionists," the Basque is outraged. The conversation is interrupted. The navigator arrives and shows on the map the location of the aircraft. Burgos remains on the right, and the Iberian mountains are visible on the left. We fly almost east and go around dangerous places. Time is surprisingly slow. Everyone is busy with his thoughts. Khativa takes out a flask and offers to drink cognac. Everyone agrees and is a little distracted. The navigator appears again and pointing to the map repeats loudly several times: "Soria, Soria." A significant part of the path is already behind. Now on the left is the valley and on the right is the Cordillera. In half an hour we will be over republican territory. For vitality, we repeat the procedure with cognac, eating some ham. So a week ago it was on the "Libertad's" bridge, when passing El Ferrol an enemy plane found us. Instead of a sober assessment about a possible meeting with the rebel ships everyone deliberately loudly talked about extraneous things. But we will not judge strictly. As everything is relative including danger. When the navigator announced that we were flying over our territory and the pilot abruptly turned to the right, then to the left, he went down, everyone came to life as if the flight had already been completed. Meanwhile, we just passed the front line in the Guadalajara region. Madrid was literally just around the corner, and the small hills in its vicinity were clearly visible under the wing of the plane. At the Alcala de Henares airdrome we warmly said goodbye to the passengers as if old acquaintances, despite the fact that we spent together only two hours of an unusual air route over enemy territory.

On the Valencia highway we drove into Madrid. The embassy representative who met me said that Toledo was occupied by rebels. Decisive fierce fighting was approaching on the approaches to the capital.

The Squadron Returns

From the very first days of the squadron's stay in the North the ships encountered certain difficulties: lack of fuel, cramped basing and lack of anti-aircraft cover during raids by enemy aircraft. Ships could only rely on their own anti-aircraft guns. There were no fighter aircraft either and enemy aircraft flew with impunity day and night.

The appearance at the end of September in Gibraltar and then in the Mediterranean Sea of the rebel cruisers, the sinking of the destroyer "Ferrandez" and the shelling of some ports of the Mediterranean Sea where the main communications of the Republicans now went, all this forced the government to reconsider the issue of basing the fleet. The erroneousness of dispatching North was realized and the ships, without having undergone any major operations there, began to prepare for the reverse passage.

In early October, the fleet received orders to return to the Mediterranean Sea.

The situation for the return passage of the fleet was much more complicated than during the journey to the North. At the end of September, enemy cruisers were in the Gibraltar area and, based on Cadiz and Ceuta, represented a significant danger to the republican squadron with a slow-moving battleship.

Now there was no chance of a sudden passage through Gibraltar. The rebel fleet was able to prepare and strike, choosing a place and time favorable to it. The enemy also had the opportunity to use aviation against republican ships throughout their course.

The personnel of the ships were ready for the return passage, since everyone understood that a fleet was needed in the Mediterranean Sea.

Leaving the destroyer "Diez" in the North, which was later joined by the "Ciscar" destroyer, the Republican fleet left the Cantabrian Sea on 10 October and headed west. It was once again to pass by the main base of the rebels - El Ferrol, then, turning south, break through Gibraltar to the Mediterranean Sea.

The enemy conducted more intensive reconnaissance and rebel aircraft often appeared over ships.

On October 12, republican ships approached Gibraltar and expected decisive action from the fascists. It seemed unbelievable that they would not try to strike at the republican squadron. It was assumed that this blow would be inflicted in the area of the strait, when the ships could not maneuver.

In the afternoon, indeed, air raids followed, although to no avail, as the planes bombed from a very high altitude. The fire on planes was opened primarily by the cruisers "Libertad" and "Cervantes" which had stronger anti-aircraft weapons, while the rest of the ships, increasing their speed, maneuvered during the bombing.

The rebel cruisers did not dare to go to sea and did not enter the battle with the republican fleet. On republican ships, it was speculated that apparently distrust of the crews kept the rebel command from trying to attack the squadron; however the true reasons for the enemy's passivity at sea remained unknown.[35]

The squadron passed the Strait of Gibraltar at dawn being at greater risk than when moving to the North. Ship handling was not always competent and sometimes only chance helped to avoid collisions between their ships.

Everyone was relieved when Ceuta remained behind the stern and the squadron entered the Mediterranean Sea.

On October 13, 1936, cruisers and destroyers headed to Cartagena, and the battleship "Jaime I" to Almeria.

The personnel and equipment needed at least a short-term pause at base.

Thus ended the bold in conception but unwarranted from a military point of view dispatch of the republican fleet to the North and back.

The information from the fronts on the day the ships arrived in Cartagena was disappointing - the African rebel forces moving

---

[35] It is now known the cruiser Canarias was off Casablanca blockading a republican vessel and returned to the Straits of Gibraltar after the fleet passed.

towards Madrid were at a distance of 50 kilometers from the capital.

Instead of North to South

Having reported on the situation in the North, I asked permission from M. I. Rosenberg and V. E. Gorev to return back as soon as possible.

There was nothing for a sailor to do in Madrid. After the first flight through Franco territory now it no longer seemed so dangerous. Waiting for the opportunity, I continued to get acquainted with the situation at sea, studied Spanish and in the evenings I talked with our journalists who were best acquainted with the situation on the fronts and in the rear. So a week passed, and circumstances changed when the opportunity arose to fly by plane to Bilbao or Santander. I was invited by our ambassador and said that in the coming days, from the Soviet Union, would arrive transports with weapons and I have the obligation to take them to Cartagena or to another port. "This is an order from Moscow," he added. "What about a trip to the squadron?" I asked Vladimir Efimovich Gorev, as my immediate military commander.

"You know that the squadron will not leave from today or tomorrow, but meeting the first transport is perhaps more important," he replied. It so happened that instead of the North I had to go South.

As it turned out a little later, the first transport with weapons was already loaded in Sevastopol and was preparing to leave for Spain. Then naturally the question arose where is it better to unload it? There was no choice. The only military port protected from the sea and from the air was Cartagena. In addition, one could count on at least some secrecy. My departure to Cartagena coincidentally coincided with the day the squadron left the Cantabrian Sea for the Mediterranean.

Fleet operations in Cantabrico

I still could not get to the North, but I did not lose interest in the events on that sector of the front both on land and at sea. What happened in Cantabrico in the following months? The stubborn struggle on land and at sea continued, but much remained difficult to do with the remaining destroyer and submarine in Bilbao. And the command of the fleet in the person of Navarro turned out to be unreliable. (In the spring of 1937 Navarro fled to France, and from there he moved to Franco.) Minister of the Sea Prieto demanded, if possible, to strengthen this section of the front by ships. Therefore, at different times they were sent to Bilbao, where the republican fleet, another destroyer and two submarines were concentrated. At the same time, measures were taken to strengthen the shore with batteries, minefields were set up, and sea surveillance was intensified. So, by the spring of 1937, two new destroyers were operating in Cantabrico: "Jose Luis Diez" and "Ciscar" and one small old destroyer "No. 3". Three submarines: "C-2", "C-4" and "C-6" also concentrated in the North. A significant number of small ships and boats were adapted for sentry service, trawling and guarding harbors. In addition to Bilbao, Santander and Gijon, small bays such as Castro Urdiales, Santonya and Aviles were inspected and equipped for the temporary anchoring of ships and small transports. Valentino Fuentes was assigned to command the fleet in Cantabrico. One day I was with Prieto in Valencia and found out that Fuentes was urgently leaving for Bilbao. I wanted to see him before leaving to agree on the use of several Soviet volunteers sent to his disposal. Returning to Cartagena and driving through Denia, my driver accidentally drew attention to a naval car near the hotel. It turned out that Don Valentino, having already handed over the destroyer "Lepanto", was heading to Valencia to come to the Minister and receive a new appointment. I spoke about Fuentes above and I will not repeat myself. I just want to note in what difficult situation he had to work in a new place and what a strong character was required for this, and Fuentes unfortunately did not have it. We met with him in a restaurant and he shouted loudly: "Hombre! Ola, don Nicholas," he invited me to join him. He was in a great mood. He was impressed by his appointment of the Cantabrico fleet command directly from the post of destroyer commander where he sat for a very long time. I congratulated him on his high appointment, and we drank for success. "Viva la Republica!" Valentino answered loudly, attracting the attention of other visitors. It seems to me that at that moment he wanted everyone to see and know that it was he, don Valentino, who was going to the crucial part of the struggle and not as anyone, but as the fleet commander. When I started a conversation about the "Companeros Ruso" who will help him, he stood up, shook my hand and, shaking it tightly, began to assure me of the sincerity of my feelings towards all Russians, and it seems to me in this case he did not deceive. When we went outside the cars were ready to continue the trip: one to the north - to Valencia, and the other to the south - to Cartagena. I wished a good journey to the new commander. After that, I never saw Fuentes again. The situation more and more attracted me to Cartagena. I learned about the actions of ships in the North from my volunteers (while I was in Spain) and from the press when I returned to my homeland. Events there developed as follows.

Until May 1937, both on land and at sea it was relatively calm. After the defeat near Guadalajara, the rebels decided to slow down the general offensive on the Central Front and shift the center of gravity of the struggle to the northern provinces. One of the large rebel ships[36] was constantly off the coast, with the aim of blocking the Republicans, preventing the supply of weapons and food and shelling of cities. Republican ships were put in order, in anticipation of decisive battles and carried out small operations to help the land flank and to ensure the movement of transports and the protection of the coast.

The war at sea on this site took place in peculiar conditions. In addition to the warring parties, there were ships of neutral states - England, France and Germany. The neutrality of the Germans has long been a fiction. England and France at this stage of the struggle formally remained neutral but did not want to help the Republicans and thereby actually played into the hands of the rebels. Many transports with goods and with the evacuated population went under the English or French flag. These ships were safe in Spanish territorial waters and on leaving they were immediately taken under guard by warships whose flag fluttered at their stern. These were as rule English or French ships. The Republican fleet was able to conduct them or meet them at the border of territorial waters and the rebel ships did not dare to approach the shore so close. On this basis there were many tense episodes. It turns out, say, bound from Bilbao or Gijon an English transport with children and civilians, and the rebel cruiser "Cervera", ignoring the flag, is trying to take it to its port, threatening people with violence. Then a destroyer, cruiser, and sometimes the battleship of the English fleet appear on the scene with the demand to free the captured ship. In some cases, it came to pointing the guns at the rebel cruiser and sending an ultimatum to release of the captured ship. Observing "neutrality", the British and French transports did not carry weapons, but only food or fuel. Weapons were delivered exclusively on Spanish or chartered merchant ships. For example in early June 1937 a transport from South America arrived in Bilbao on which there were 10000 rifles, automatic rifles and machine guns. The destroyer that met him distracted the cruiser "Cervera" upon themselves while the transport, using the darkness, reached port.

The transport "Habana" was involved in the evacuation of Spanish children and women from Bilbao. The heroic behavior of the crew of this ship is known. Eyewitnesses told me what dramatic episodes were accompanied by each loading of people in Bilbao,

---

[36] The battleship Espana.

especially in early June 1937 ... when the rebels were already advancing on the city and bombing the port.

I remember the operation to evacuate women and children to the Soviet Union. It happened in May and June. Several steamers such as "Ciudad do Cadiz", small but comfortable for passengers, took Spanish children to Cartagena and set sail. First of all, these were the children of the dead refugees who took refuge from the atrocities of the fascists. They were escorted by Spanish comrades and the ship crew met them with great attention. I did not want to send such ships without security; they were most afraid of falling into Franco's clutches. We used this method: either timed their exit to the next exit of the squadron at sea, or the commander provided a destroyer, who escorted steamers with children to the territorial waters of Africa in the Algeria region. Even in those days when the danger of being captured by Franco's or Italian warships increased, we did not have a single case with a sad outcome.

But back to the Basque Country.

From June to the end of October there was a tragedy in the unequal struggle of the brave but poorly armed and hastily trained fighters for the Republic against numerous German-Italian aircraft and armored units advancing from east to west. The Basque corps, until then stubbornly resisting, could not stand it, and the enemy in the first half of June began fighting in the immediate vicinity of Bilbao, breaking through Cinturon - this defense belt, on which the fate of the capital of Basque Country depended. Hundreds of thousands of refugees filled the highway leading to Santander those weeks. By this time, the rebels had concentrated the entire fleet on this sector of the front and established a close blockade of ports. The Republican fleet was forced to relocate to Santander, where there were not enough supplies of fuel, anti-aircraft defenses or repair facilities. The "Diez" and "Ciscar", to the best of their ability, guarded the shore from a landing. They went out to sea more than once, but during the day they were weak to fight the large surface ships of the rebels, and at night they were busy escorting numerous ships moving to Santander and Gijon. Despite the superiority of the rebel forces, the destroyers managed to repeatedly break the blockade and ensure the arrival of transports with food and fuel. Sometimes they used a ruse: one destroyer joined battle with the cruiser, dragging him along, and the other meanwhile escorted transport to the port. If the ships did not yet have the proper order and ability to competently fight, then the courage of the personnel cannot be denied. Submarines bravely, although unsuccessfully, repeatedly attacked the "Cervera".

After the fall of Bilbao, the situation became critical. Santander was not strengthened properly and resistance weakened every day. At the end of August we had to evacuate that as well. Asturias, with the small town of Gijon, had to take on the defense of the last frontiers of the republican North. Ships operated until the last opportunity. Damaged “Ciscar” had to be scuttled. “Jose Luis Diez” at the last moment left for Le Havre. Submarines “C-2” and “C-4” left for Saint-Nazaire and Bordeaux. Hundreds of small ships and fishing vessels at their own risk and fear broke through the blockade and went to sea. Foreign transports, ignoring danger, until the last moment called at Gijon, took on board the population and disparate groups of fighters and took them to France. On October 20-21, 1937 the northern provinces passed into the hands of the rebels.

At the Distant Meridian

Closer to the fleet

In Cartagena

On October 10, 1936 I flew to Cartagena. At the request of the republican government to the Soviet Union it was planned to send the first consignments of military cargo. The day before I visited I. Prieto. The conversation took place in his apartment late in the evening. Over a cup of coffee with cognac, Don Indalecio talked and, as it seemed to me, wanted to emphasize the informality of our conversation. He praised the passage of the fleet from the Mediterranean Sea to the ports of the northern provinces, although he did not offer to drink the bottle of champagne promised at the time.

By this time he was already Minister of Aviation and the Navy and sat firmly in these two chairs at the same time. Better informed than myself, he understood communications with the Soviet Union were of particular importance. The war took a protracted character. The organization of the people's army and the acquisition of international brigades depended on the receipt of weapons. Some were possible to obtain from France, Mexico and other countries, but this was clearly not enough. By this time, it had already become apparent that the basing of the fleet in Cantabrico was difficult; moreover, he performed only private tasks that did not have a decisive influence on the course of the civil war as a whole. “I made the decision to return the fleet to the Mediterranean,” Don Indalecio began the conversation. I already knew that. We exchanged opinions the return voyage would obviously be more difficult than the quick passage of the Straits of Gibraltar we did in September, and the rebel fleet was replenished with the new cruiser "Canarias", which, with a bold and initiative command, could deliver a lot of trouble to our squadron with the slow battleship "Jaime I". Prieto now considered desirable my presence in Cartagena. "I spoke with your ambassador Rosenberg, and he promised to give the necessary instructions," the minister concluded. I had no choice but to express my readiness to do everything in my power. Our ambassador really pointed out to me the need to go to Cartagena and concentrate all my attention on ensuring the acceptance of goods. When the first transport with weapons shipped out, the question of escorting it with the republican fleet was not yet raised, and organizing unloading at port did not present a great difficulty. "What about the passage of the squadron? Which I did promise to return to." I asked the ambassador. However, the departure of the ships would actually take place on the day of our conversation and there was no possibility to fly there as we did two weeks ago, over the territory of the rebels. Our adviser N.P. Anin remained with the fleet commander M. Buiza.

My flight to Cartagena was scheduled for early morning. From the Alcala de Henares airfield, located 25 kilometers from Madrid along the highway to Valencia, a scheduled plane flew to Alicante. From there to the base less than two hours by car. At the appointed hour, about 15 passengers of different nationalities were accommodated in the old "Potes". The small town of Alcala de Henares was known primarily for the fact that the writer Cervantes was born in it and now it fell to him to base fighters with Soviet volunteer pilots and protect Madrid from the Junkers and the Heinkels and Fiats that accompanied them.

We flew to the rear and many passengers were apparently satisfied. Madrid had turned into a front-line city; the rebels were preparing a powerful offensive, surrounding the capital from three sides. The Spanish population did not want to leave the city, but numerous foreigners and embassies preferred to move to the rear, spreading rumors about Franco's imminent and inevitable victory.

A strong turbulence in a mountainous area caused great trouble. It ceased as soon as the mountains passed and the plane quickly began to descend; the blue sea opened ahead. Then began usual revival of spirits before a landing. The stewardess asked everyone to fasten their belts. I was not sure if there would be a car from Cartagena, but that did not bother me anymore. To my delight, even from within plane I noticed a gray Hispanic-Suiza and my small statured, heavy set, sailor-chauffeur Martinez: it means that everything is in order. Only due to the quick descend did hurt my ears a little and after the strong turbulence everything swayed around, like on deck in a ship during a storm. Two hours drive along the beautiful coastal road, and we already passed the airfields of Los Alcazares and San Javier. Here is Cartagena. The harbor is empty and there is ordinary activity in the city. This is a situation familiar to the sailor when the fleet goes to sea and the base, temporarily orphaned, awaits its return. And here's the building with a small sign: "Vice Almirante, Héfe de la Base de Cartagena" - Vice Admiral, commander of the base of Cartagena. Antonio Ruiz welcomes me warmly and invites me to his office. He is glad for my arrival, but, as I think, not because there is a lot of useful work ahead. He did not yet know that in a few days the first Soviet tanks would pass through the city and the people would enthusiastically shout: "Viva Russia!" It is precisely these volunteers would take part in repelling the violent attacks on Madrid in early November. He is pleased to chat with a new person without thinking about high matters. I am also pleased that I have the opportunity, as long as there is no translator, to speak with him in French; I know that Antonio will willingly go with me and show the base, its ancient buildings, left over from the Romans and the Moors. All this is good and useful, but there are other issues that are more urgent and complex. Indeed, over the past two months I have managed to get to know many things, make a more complete picture of what is happening and think about the role of the republican fleet that it will have to play if the war drags on. And its end is not visible. Franco will no longer be able to quickly win even if he succeeds in capturing Madrid and it was difficult for the Republicans to count on suppressing the rebellion in a short time.

The fleet often has to carry out hard, but inconspicuous work, on whose success the land fronts depend on. This work is sometimes very obscure to a simple observer such that the fleet is reproached for its inaction, and in this case only command discipline ensures its normal combat activity. There is a rumor Japanese admiral Togo's, the Russian-Japanese warrior, house was burned down due to public perception of inactivity, he acted, waiting for a more favorable environment. It is also known when in the First World War the German fleet made separate raids on the English coast, the commander Admiral Jellicoe was sharply attacked, notwithstanding with any fleet he could not completely guarantee England against night raids by the enemy fleet. Public opinion does not always understand the intricacies of war at sea. So it was in this case. The Republican squadron was faced with hard, unsung work. Looking a little ahead, we can recall how from November 1936 until March 1937 the squadron continuously went to sea and protected the necessary cargoes for the army, and the apparent inactivity of the fleet caused complaints, and in April the fleet conducted a special operation to please public opinion; using even the old battleship "Jaime I" in it with the risk of losing the ship without a great combat effect. There is nothing to be done!

"Apparently, you will have to take more than one transport into Cartagena and therefore count on a long stay there," our military attaché Gorev advised me before leaving. It became obvious that the republican fleet and base will have to be prepared for the fight against the rebels against surface ships, submarines and aircraft. On the day of my arrival, the fleet was still in Cantabrico and the base was calm. For several weeks there had not been a single raid and all the activities of the rebel squadron were concentrated in the Gibraltar area. Even the blackout was no longer observed and in the evenings, as in peacetime, all the restaurants and cafes were open on the main street of Calla Mayor. Visitors at tables laid directly on the pavement lingered until late in the evening, and only the restless "pistoleros" (mainly anarchists) sometimes violated the silence of the city. In addition to goods, volunteers were expected to arrive. I was looking forward to this reinforcement of volunteer sailors. It was hard to work alone. I used the free days for a detailed study of the base: its coastal artillery, anti-aircraft defenses, anti-submarine defenses and the readiness of the port to accept cargo. Without a base commander, you won't do anything with traffic, and stirring up Don Antonio was my first and main task.

"The war seems to be dragging on, and the government is doing the right thing in seriously organizing the people's army," I began then to explain the role of the fleet and the base and especially the tasks that fall to its lot.

"No puede ser" (it cannot be), Ruiz answered me with frivolity and excessive optimism, immediately destroying my plans. "The rebellion either succeeds immediately or is doomed to failure," he philosophized. According to him, the republic is very popular among the people and the fascists have no support. "But military power is still on their side," I objected and reminded him of the fall of Badajoz, Talavera and the advance of the rebels towards Madrid. "I am also confident in the victory of the Republicans, but for this we must persistently and in an orderly manner wage a long struggle, if necessary," I campaigned, trying to set it up in a businesslike manner. I even wanted to frighten don Antonio a little with the victory of the rebels but he reacted to this rather calmly. Subsequently, I happened to observe the behavior of many officers who remained on the side of the government and only a few of them were resolutely and completely against Franco. Most hesitated, doubted, and for a long time could not make a definite choice. Too much connected them with the old system. In addition, Antonio Ruiz was an officer who could only go with the flow and was not at all adapted to the turbulent events that Spain was going through.

"The army will need a lot of weapons and no matter where we get it from it will probably arrive by sea and will certainly go to Cartagena," I made clear hints. Antonio perked up. He guessed that my arrival was connected with some plans and carefully listened to what I would say next.

"Let's see on the map how this may look in practice," I said and invited Antonio Ruiz to the table where a whole stack of sea maps of the Mediterranean lay. Having chosen one of them on which the whole Mediterranean Sea was located I set about discussing the possible communications and the dangers that might arise when moving transports at sea and being at the base.

At first glance, it became clear two lines of communications - from the Soviet Union and from France – would acquire the greatest significance if the war assumed a protracted nature. What cargoes will come from France, I had no idea. The "non-intervention" committee by this time had even stopped sending equipment on republican ships and aircraft from England and France and only a few goods were purchased there after "good lubrication". The communication line went from the Black Sea through the Dardanelles, crossed the Mediterranean Sea from east to west, with the final port in Spain. This line acquired special significance. Transports going along this path will have to pass near Malta, Sicily, Bizerta and make their way along the African coast to the Iberian Peninsula. The rebels owned Gibraltar and the island of Mallorca, near Cartagena, and therefore had the opportunity to intercept these transports in the absence of a republican convoy. Ruiz listened to me carefully. As if wishing to anticipate his question, I said that the republican fleet will soon return to Cartagena and will certainly cope with this task. "Muy bueno, Muy bueno" (very good, very good), he was delighted and asked if I knew about the date the ships were to return. Promising him to immediately tell everything that I would know, I turned to questions about the readiness of the base in anticipation of a large flow of military cargo. "If not a secret, tell me what will your base do if several transports arrive here?" I turned to Antonio carefully, wanting not to offend his pride.

"Hombre!" Exclaimed my interlocutor and immediately ordered to bring him port plans.

It turned out that the port, which had previously exported fruits and lead was not suitable for bulky goods although it had quite a few small cranes. 3-5 transports could stand and unload at the same time, if there are enough movers. The bottleneck was the means of exporting goods from the port: cars and railways. There were not enough open platforms and wagons, almost no cars were at his disposal, and all the more with trailers for transporting huge boxes with airplanes.

Antonio immediately gave orders to call the people he needed. "Call Senor Mateo for me," he ordered, trying to give a stern expression to his face. Mateo, I already knew a little. He was a colonel of the quartermaster service, in charge of all the economic affairs of the base. A tall, thin old man with sad eyes later turned out to be an indispensable worker. He enjoyed authority among all personnel. Mateo survived a personal tragedy. At the beginning of the rebellion the anarchists killed his only son when the latter tried to prevent the theft of weapons. Left alone, he devoted himself entirely to service, being devoted to a legitimate government. After serving in Cartagena for many years, he knew all the details of the economy of the base and more than resolved a most difficult situation. I didn't want to interfere with Antonio's conversation with his subordinate and therefore, when Mateo entered the office, I stepped aside, looking at the paintings hanging on the wall from the time of the galley fleet. "Perhaps one of these galleys came out of Cartagena and Cervantes," I thought.

Having received some orders, Mateo respectfully reached out and with the words: "sir senior," he put his hand to the headdress.

We continued the conversation.

I knew that Don Antonio was a follower of Prieto, and therefore found it useful to mention my conversation with the Minister and convey to him what importance he attaches to the latest transport operations. But it was not in the customs of the base, even during the war, that it worked until late in the evening, and Antonio beamed when a marinero reported: "Comida sirvido" (lunch is served). I was also pleased, as we agreed early in the morning to begin familiarizing ourselves with the situation on the spot.

We started our detour with the arsenal, which, together with ship repair workshops and docks was the main naval object of the naval base. The Arsenal was actually a small harbor which used to be sufficient in size for the entire galley fleet and at present it only acceptable for destroyers and submarines. Larger ships before the rebellion were not based on Cartagena and now they were standing in the Mall of Curro, a large harbor built not so long ago to increase the port's water area. Around the small harbor basin where we drove there were buildings of workshops, warehouses and administration. We were met by the head of the arsenal Patricio Casado, the secretary of the arsenal Melia and the chairman of the local trade union of the arsenal workers committee (aka head of the workshop corps) Artajo. This peculiar triangle, speaking our language, worked amicably and did a lot to ensure the combat activity of the fleet in subsequent years. There were no large ships at the harbor moorings. Only at the workshops were two unfinished destroyers - "Jorge Juan" and "Ulloa" - and two recently started large submarines of the "D" type. Now it was not up to them and all the means of the base were thrown to repair warships.

Here I had the opportunity to learn interesting facts of the life of the republican fleet. Constructora Naval, the company that built the Spanish fleet, turned out to be the property of an Anglo-Spanish joint-stock company was controlled from London and depended on British supplies. Taking part in the construction of destroyers and cruisers, the British have long controlled the construction of the Spanish fleet. Now this was expressed in a delay in the supply of the most necessary equipment and materials which were already paid for. Water for military and domestic needs - and that was in the hands of the British. Received from Murcia, it went through the water supply which was maintained by an English company and the main owner was none other than the British consul in Cartagena. Accidentally or not, Cartagena suddenly began to receive little water and the pumps on which it supposedly depended required replacement. They were expected from London but the same notorious "non-intervention" committee temporarily prohibited their sending to Spain. The Anglo-Spanish company, having its branches in Ferrol, which was in the hands of the rebels, and in the Republican Cartagena, played two tables. In Ferrol, she completed the construction of the cruiser Canarias and Baleares, and in Cartagena the destroyers “Jorge Juan” and “Ulloa”. "Designer Naval" remained "out of politics" when it came to profits. It is curious that not only the British consul depended on this company, but Don Patricio - the head of the arsenal - considered himself more subordinate to London than Madrid. If we add to this that the telephone company - Union Telefonika - was also dependent on foreigners, and the communication line often passed through the territory occupied by the rebels, then a complete picture of the difficult situation that Cartagena was experiencing from a military point of view would be compiled.

“Connect me with London,” Casado told the secretary, completely unaware of the secrecy of the negotiations.

“This moment,” she answered and returning reported she had requested a connection for 1600 hours however the telephone operator could not vouch regarding certainty, since the connection was unreliable.

It was difficult to change these orders and our task was to do everything possible to help the Republicans in their struggle under these conditions. At the moment described I was primarily interested in the safety of arriving cargo.

It seemed to me that an arsenal surrounded by a high wall would be the best place for unloading transports. Patricio pointed me to the pier where destroyers usually stood, now empty. We decided to drive to the place and see. It was here that a few days later the “Komsomol” unloaded delivering the first tanks and a group of volunteers.

I asked if the authorities of the arsenal intended to complete the construction of “Jorge Juan” and “Ulloa”, at first glance almost ready destroyers. “It all depends on supplies from England,” said Casado, “now work has been curtailed, since the workshops are loaded with repairing warships." Then it was not believed the civil war would drag on for three years and much would change in the arsenal itself, and these destroyers - completed - would still take part in the battle.

Saying goodbye to Patricio, Antonio and I drove out onto a 15-inch coastal battery. Two of these batteries, from ship-mounted guns, once built for the British battleships, were installed to defend the base. The road went in zigzags, steeply climbing up. How much work was required to drag heavy turrets onto such a mountain, but the battery became almost invulnerable from the sea, and not a single ship dared to approach it.[37]

"Mountains without greenery," the Spaniards say about Cartagena. There is really little green here. In early spring, it appears on the slopes of the mountains but the hot sun in June-July burns it ruthlessly and only in certain places does the greenery remain year round. The sea appears in sharp contrast with the gray city and its environs. Blue, it merges on the horizon with the same sky, and you can look at it for hours, discussing the transience of human existence. "Mar Sin Pescado" (the sea without fish), say the inhabitants of Cartagena and are content with mollusks, making excellent arroz de la pola (pilaf made from rice and edible mollusks). When we stopped at the battery, the artillery chief, Captain Mira, reported to Antonio and introduced the battery commander. "Sea without fish" was calm. The mirror-like calm light surface only darkened in places where the wind caused small ripples. The city seemed to merge with the surrounding monotonous gray area, and only the bay shone, sandwiched among the high mountains. From here it was easiest to appreciate ancient New Carthage as a happy combination of deep-sea approaches and harbors sheltered from all winds. This made the port very valuable in the era of the galley fleet and today with aviation and submarines.

The commander ordered to sound the alert.

---

[37] Bateria de Cenizas

The newly installed coastal defense battery turrets were set in motion and the cannon barrels lifted upward spoke of the strength and combat effectiveness of the coastal defense.

Captain Mira offered to go to the command post and it seems was going to make a test shot, taking advantage of the presence of the base commander. "I'm sure even without this that in case of need the battery will shoot well," I told Antonio who looked at me. Such a need came for her a few months later, when the "Canarias" and "Baleares", in pursuit of a republican destroyer, approached Cartagena, and Cenizas' battery saved the destroyer, firing several large-caliber shells. With these coastal defense in Cartagena was actually safe. Some batteries however were so outdated that they could not be taken into account but enemy cruisers did not pose a serious threat to them.

I was interested in one of the 3-inch anti-aircraft batteries located nearby. An air attack posed a real threat. The four-gun anti-aircraft battery was clearly not modern. The old guns were not equipped with detection tools and control devices were designed to fire using visual pointing. Such batteries could not reflect air raids at night. I noted for myself the need to strengthen air defense systems, but in order not to offend the personnel, I praised the commander. "A Sus ordenes" (at your service),"he answered cheerfully and we left.

I did not know then Cartagena would be my permanent place of residence and I would have to visit Valencia where the government moved-to only for reports in my spare time from operations.

The actions of the rebel fleet

The first rebel ships that launched combat operations against the Republicans were the cruiser "Cervera" and the destroyer "Velasco". They were used in the North to block the coast and shell Bilbao, Santander and Gijon.

In late August, the cruiser "Canarias" entered service. Together with the cruiser "Cervera", he already took part in shelling the coast when ground units of the rebels attacked Irun and San Sebastian.

Since the Republicans did not have any ships in the North except the old destroyer "No. 3", rebel ships acted there with impunity. In September alone, three Republican submarines appeared in the North which limited the actions of the rebel ships off the coast but they could not seriously counteract them.

On September 19, 1936, the rebel destroyer "Velasco" sank the Republican "B-5" submarine near Santander. Caught on the surface and for some reason unable to dive, the approaching destroyer destroyed it near the shore and shot the crew trying to escape.

At the end of September, when the republican fleet moved to the North, the rebel "Canarias" and "Cervera" relocated to Cadiz. On September 29, they attacked Republican patrol ships, sinking the destroyer "Ferrandez" and causing damage to the destroyer "Gravina". Then the rebel cruisers moved to Cadiz and Ceuta, began firing at vulnerable ports and towns of the republican coast. As a result of one such shelling oil tanks in Almeria were destroyed which limited the already small fuel reserves of the Republicans.

On October 13, the republican fleet returned to Cartagena. The rebels in the North left the battleship "Espana" and the destroyer "Velasco" to try and block the republican coast.

So ended 1936. After active and intense months of struggle (July - October), there was a temporary lull as the fleet of both sides required repairs.

Franco's first major operation in 1937 was an attack on Malaga. The rebel fleet, significantly strengthened by that time, actively participated in this operation.

The three rebel cruisers – "Canarias", "Baleares" (which went into operation in January 1937), and "Cervera", based at Ceuta and Cadiz, actively assisted the ground forces during the offensive on Malaga.

On January 14, 1937, all 3 cruisers with 2 gunboats "Castille" and "Lauria", accompanied by minesweepers, fired on Estepona at the approaches to Malaga from the south and then repeatedly appeared in the area, supporting their land front.

On February 7 and 8, when decisive battles for Malaga took place, the rebel cruisers continuously fired on the city and its approaches. It should be noted the good interaction in this operation of rebel land, air and sea forces.

On the evening of February 8, Malaga was captured by the Francoists. Attempts by the republican fleet to counteract the rebel cruisers were unsuccessful as German ships threatened maneuver nearby and Italian cruisers repeatedly appeared on the horizon, misleading the command of the republican fleet and distracting from the actual enemy ships.

The active participation of Germany and Italy in the war on the side of Franco was not a secret from the first days of the rebellion. This assistance increased after the rebel plan failed to quickly end the Republican government. Some interesting data, as the former government of fascist Italy acknowledged, during the Civil War the rebels received: 2000 guns, 10000 assault rifles, 240000 rifles, 324 million different cartridges, 8 million artillery shells, 12000 trucks , 700 tanks, 800 aircraft, 17000 bombs, 2 submarines and 4 destroyers.[38] I think these figures if they suffer from inaccuracy are in the direction of an understatement.

By participating in the development of the plan of rebellion, fascist Germany was able to take the side of the rebels from the very first days of the struggle. By the beginning of the revolt off the coast of Spain there was already a German squadron

---

[38] Author's Note. "Military History Journal" No. 12, 1960, pp. 118–120.

consisting of two battleships of the Deutschland type, three cruisers of the Nuremberg type and a destroyer flotilla.

Italian aviation transferred troops from Morocco to the peninsula, and Hitler already in August provided Franco with transport planes, Junkers 52 and fighters. When it became clear that a long struggle was ahead, Hitler ordered the creation of a special aviation Condor legion. In August 1938, there were already more than 400 fighters and about 300 bombers in this legion.

I have been thinking a lot: was it possible to keep Malaga? Without a doubt it was possible. Malaga was attacked by the same Mussolini units that were later beaten at Guadalajara. The offensive was known at least a month in advance. Defense, as you know, always requires forces two, three, or even four times less than an offensive. So, without weakening the capital's defense ring to unreasonable sizes, it was easy to draw some of the forces in the south.

Unfortunately not enough attention was paid to the southern front. General Asensio, Caballero's chief adviser and actual military operations leader was not up to par. The Communist Party strongly demanded the Perhaps as early as January it was necessary to go beyond the report of Bolivar, the Commissar in Malaga, to L. Caballero and raise the issue with the Government with more determination than that. As for the republican fleet, he freely could allocate part of the ships and base them at Malaga. Positioning two or three destroyers there would make it difficult for the rebel squadron to approach the city. But the central government should have done this. Ground defense, aviation and the navy under a single command could protect Malaga. And since this did not require great operational knowledge it is difficult to say where negligence and carelessness end and betrayal begins.

After the capture of Malaga the rebels reliably controlled the Strait of Gibraltar and the Republican fleet was forced to limit the activities of their ships to the Mediterranean region east of Malaga.

Having finished the offensive in the Andalusian direction, the rebels intensified in the North.

In April, an offensive is launched on Bilbao. Two cruisers and a rebel battleship constantly operated there, blocking the coast and shelling ports and coastal roads. By this time, the Republicans had two destroyers in the North – "Diez" and "Ciscar" - and three submarines. It was difficult for this unit of the republican fleet to do anything against the cruisers and the battleship of the rebels. But even in these difficult conditions, they were successful. So, in August 1937, the tanker "Valletta" with 8500 tons of fuel and several transports with food were on the approaches to Gijon and there were also transports with food and other cargo near Santander; Republican sailors undertook a risky operation: two destroyers simultaneously went to sea, deciding to engage in an unequal battle with the cruiser, in order to divert the enemy's attention to one of the destroyers, giving the other the opportunity to escort the transports to the republican port. It was August 11th; that day, the tanker "Valletta" and several food transports were escorted to Gijon and Santander. However, the help received by the sea was not enough. The preponderance of forces was on the side of the rebels and the actions of the ships were limited to protecting the coast from shelling and landing. From the beginning of March until the end of October when Gijon fell there is a stubborn struggle for the North. During this period, the destroyer "Ciscar" after heavy damage fell into the hands of the rebels, the submarine "C-6" was destroyed. The destroyer "Diez" went to France, repaired with great difficulty and wrote another glorious page in the fight when trying to break into Cartagena. After an unequal battle, he was interned in Gibraltar. Submarine "C-6" operated with exceptional courage until the last days while being in northern Spain. Unable to go out to sea due to damage the boat was blown up by the crew to a loud: "Viva la Republica!" This boat is under the command of the communist commander NP P. Egipko (Matisse) and commissar Paolo several times attacked fascist ships. Matisse later recalled with bitterness how difficult it is to fight when the boats have one set of poorly functioning Fiume torpedoes purchased by the government back in 1928. Interestingly, commander Matisse and crew were later on another boat "C-2" and reached Cartagena after a long ordeal and repairs in France. Commissar Paolo died heroically in the struggle for the Republic.

In April 1937 a mine detonation sank the battleship "Espana" rebels died. Occupying the majority of bases, the rebels, depending on the situation concentrated their cruisers in the North or the Mediterranean Sea.

The main task of the rebel cruisers in the Mediterranean was to fight on communications coming from the East along the shores of North Africa to the ports of the Republicans, primarily to Cartagena.

So, on September 6 and 7, 1937, a convoy of republican transports collided with the cruisers "Canarias" and "Baleares", as a result of which the transports were forced to return to Algeria.

After the fall of Gijon, in October 1937 and the liquidation of the Northern Front all naval operations on both sides concentrated in the Mediterranean Sea. By this time the forces of the rebels exceeded the forces of the Republicans, nevertheless the Republicans boldly went to sea and at night they searched for the enemy using their destroyers to strike him. In March 1938, there was a clash of Republicans and rebels in the area of Cape Palos ending with the sinking of the squadron flagship cruiser "Baleares". We will return to this battle, it was the last major clash at sea.

Having settled in Cartagena I often went to Madrid to solve various problems and then to Valencia where the Spanish government and our embassy moved. I had to maintain close contact with Minister of the Navy I. Prieto and G. M. Stern (Grigorovich), who led our volunteers on land fronts. I would rather have once again personally reported to the Minister of the Navy than to use a radio or telephone for confidential conversations. The Minister of the Sea knew what important cargoes went by sea from the Soviet Union and instructed the fleet commander to timely fulfill all my requests for meeting, unloading transports.

G. M. Stern was extremely interested in the timely delivery of goods. He understood that it was not appropriate to talk on the phone about the receipt of weapons, but still sometimes called me in Cartagena to find out how I was doing.

- When will you come to us in Valencia? He usually began his conversation. This meant: by what time would the transport be unloaded?

"I'll leave tomorrow," I answered him, trying to quickly end the dangerous conversation. We worked very well, G.M. Stern and I, and in a word, understood each other. After returning from Spain, fate brought us together again in the Far East, and I still have the best memories of our joint work.

In connection with trips to Madrid and Valencia, I recall several episodes.

Returning once from Madrid, I could not resist the temptation to drive along the "path" of Don Quixote.

"This is the same molino de viento (windmill)," says my interlocutor, a connoisseur of Spanish attractions, Juan Lopez, "where he stayed and talked with the beautiful Dulcinéa, hidalgo Don Quixote." It was a road that all admirers of the talent of Cervantes, who was not only a famous Spanish writer but also a brave soldier who participated in the battle of Lepanto, considered it obligatory to drive. One evening, when low cloud cover and light rain (which is not so often in Cartagena) did not portend raids a conversation about Cervantes entered the dining room of the base commander. Antonio Ruiz said that the famous Spanish writer participated in the war with the Turks and distinguished himself in the famous Lepantine battle, while he was still a very young platoon commander in one of the Spanish galleys. He was not a sailor in the full sense of the word. While in Italy Cervantes enlisted as a soldier in the Spanish units forming there and with them ended up in galleys. Then there was no long-range artillery, heavy oars served as engines, and soldiers fought. It was, in modern terms, the marines armed with arquebuses.

We imagined how 210 galleys and 66 galliots of the Turks clashed with 201 galleys and 6 galleasses of Christians, coming close to each other. The physical strength and personal courage of the naval soldiers decided the outcome of the battle.

On one of the galleys the sick, eaten by fever, immortal creator of Don Quixote M. Cervantes fought bravely. He was in the most dangerous places of the battle and was injured four times after which he spent many months in a hospital in Messina. True, not for this he is put on a par with Pushkin, Shakespeare or Goethe, but this deserves to be noted in his biography.

The battle of Lepanto was the last victory of the Spanish fleet. It was the year 1571. The 16th century, brilliant for Spain, was departing and with it the fleet's former power was also a thing of the past. The Spaniards love to remember the battle at Cape Lepanto and are reluctant to talk about the “Invincible Armada”, 17 years later it was defeated which was a harbinger of the imminent decline of the entire “empire of the setting sun." For Miguel Saavedra Cervantes the battle of Lepanto was also the last military success. Despite serious injuries he, however, participated in the unsuccessful Tunisian expedition with the Spaniards. Two years later, on a campaign in Algeria Cervantes was captured where he was repeatedly threatened with death. Only in 1580 was he bought back. Returning to his homeland he spends three more years in sea voyages but military service becomes impossible for him. Wanderings in search of a piece of bread begin. Having lost his hand he was forced to leave military service and engage in literary activities. After everything I heard about Cervantes the soldier I wanted to repeat the fictitious path of the heroes of the famous novel and get acquainted with the monuments of that time in more detail but the situation did not allow this.

Unfortunately, it was not possible to establish the place where the remains of Cervantes rested, a lot of monuments were erected to him and Madrid is considered the best. How many years have passed since Cervantes observed and described Spain but even during the Civil War the features of the good and naive Don Quixote could be found among many Spaniards. This is especially true in the early days of the Civil War when the fierce struggle against the rebels was incomprehensibly combined with an amazingly naive credulity towards their irreconcilable enemies.

In Valencia I met many friends. The Metropol Hotel gathered in those days writers, journalists and various political figures from around the world. There you could meet A. Tolstoy and I. Ehrenburg, E. Hemingway and Mate Zalka.

Usually we went at night - it was faster and safer. Our driver, the sailor Ricardo, loved a fast ride. He drove the car at a speed of 90-100 kilometers per hour. Once, on the way to Valencia, an oncoming car hit us into a ditch. They got off with light injuries, but after a couple of months I still limped on my left foot. The most interesting thing was Belgian journalists were in the oncoming car whom I first met in September on our steamer "Neva" in Alicante. Saying goodbye then, they assured us that we would definitely see each other again. "Well, here we are," one of them said smiling when I was already sitting near the car. They explained the collision by the carelessness of their driver: "We warned him more than once, but he chatted and did not look forward." It didn't make me feel any better.

I still remember the unexpected flight from Valencia to Barcelona in March or April 1937. In Barcelona I reccived the opportunity to spend two evenings with our consul V. A. Antonov-Ovseenko whom I already knew. He met me with M. Koltsov who had just arrived from the Aragon Front. We sat and remembered the Motherland. I spoke about the situation in Cartagena and on republican ships. Mikhail Efimovich did not lose hope of visiting and even sailing in the ships of the Spanish fleet. These were my last conversations with both V. A. Antonov-Ovseenko and M. Koltsov. Their fate is tragic.[39] It was 1937.

We will return, however, to Cartagena.

Provision of shipping from the USSR to Spain

Undoubtedly, the most valuable contribution of the Republican fleet to the fight against the rebels was the escorting shipping to and from Spain. Maintaining the communications of the Republic with the outside world, important under any conditions, acquired special significance in 1936–1937. due to some specific circumstances. When rebellion and war broke out in Spain, England and France reached an agreement between all European states on non-interference in Spanish affairs. The USSR, having acceded to this agreement, continued to actively defend the interests of republican Spain, exposing at the same time the true meaning of the policy of the Western powers in relation to the Spanish Republic. It soon became clear that Germany and Italy, having signed an agreement on non-intervention, brazenly violate it sending Franco weapons, military advisers and even entire armies while England, France, the USA refuse to sell even gas masks to the legitimate government of the Spanish Republic. The USSR then openly declared that it did not consider it possible to comply with the agreement to a greater extent than some of its other parties (meaning Germany and Italy) and began to sell weapons and ammunition to the Spanish Republic. Since that time the Soviet Union has become the main supplier of weapons; without this help it would be difficult for the Spanish people to continue the struggle against fascism. In the years 1936-1937.weapons were transported mainly by sea - from Soviet ports on the Black Sea to Spanish ports on the Mediterranean Sea, primarily to Cartagena. Francoists, as well as fascist Italy and Germany, sought by all possible means (including shelling, bombing and torpedoing of ships) to impede free communications between the USSR and Spain. The Republican fleet was to escort

---

[39] Both were later liquidated in the Stalinist Purges.

vessels with precious cargo to Spain under any conditions. I must say bluntly that, despite all difficulties, the republican fleet coped with this task. More than 20 major transports safely passed from the Soviet Union to Spain until September 1937 and only the active intervention of Italian ships disrupted sea communications with the East at the end of 1937. I would like to tell you something about this glorious page in the history of the republican navy.

As early as September 1936, the Soviet vessels "Neva" and "Kuban" arrived from Odessa in Alicante with food and clothing for Spanish women and children sent by Soviet unions.

In October, ships with weapons and ammunition began to arrive. The first such vessel was the "Komsomol", which delivered tanks, vehicles and a small number of guns. Like "Neva" and the "Kuban", "Komsomol" sailed without any support from the republican military authorities. Only the pilot on a boat met him to indicate the mooring spot in port. The captain of the transport was G. A. Mezentsev. On board was a group of volunteer tankers led by S. M. Krivogshein. Soon after, a large Soviet transport, "Kursk", arrived in Spain, delivering the first batch of fighter aircraft, air bombs and gasoline. Aircraft as well as tanks were necessary for the fight near Madrid.

Interesting memories are associated with the advent of these transports. When the "Komsomol" anchored in the harbor of Cartagena I went on a boat to the captain of transport. We quickly agreed with him, with tankers the matter was more complicated. S.M. Krivoshein, as senior, demanded from me documents giving him the right to dispose of the cargo. At that time neither I nor anyone else could have such documents. Only he alone knew what load and how much he delivered. I only knew that this cargo should be sent as soon as possible to Archena and then to Madrid. There was no representative of the Spanish army in Cartagena at that moment but there was no time to lose: at any time one could expect a raid from Franco's aviation. Secrecy in the conditions of Cartagena was a hopeless affair and I tried to convince the tank men to immediately begin unloading. While G. A. Mezentsev was mooring a land or rather a tank comrade whom

Krivoshein believed as a colleague in his specialty arrived in the arsenal although he also did not have a mandate.

The first batch of tanks was still in the arsenal behind a high wall and the whole city already knew and animatedly discussed this event. When Soviet tanks under their own power went through the city to Murcia and further to Archena where they had to hurriedly prepare for the heavy fighting for Madrid there was no end to the jubilation of the population.

- Viva, Russia, Viva! - shouted the crowd and threw up hats.

They rushed to Madrid where it was dear there every day and the Franco-General General Capo de Llano from Seville having heard about the arrival of a shipment of weapons to the Republicans had already threatened on the radio to destroy Cartagena and showered vulgar abuse on all the “Reds”.

Unloading was slow, but most importantly, the carelessness with which the Spaniards handled ammunition and gasoline bothered me. Boxes of shells were thrown onto the truck as if they were oranges. Smoking near gasoline barrels was the second scourge and don Antonio, the base commander, addressed the workers several times with exhortations but this did not help much. Why then, did people who are ready to give their lives for the Republic not understand basic things? It seems to me that this can be explained by the naive carelessness so characteristic of Southerners and the lack of military experience (after all, Spain has not really fought since Napoleon's time).

I had to ask the tank crews and the ship’s crew to set their own security on the dock, though to do it was not quite fitting in a foreign port. But there was no other way out; the Spaniards themselves understood this and were not offended.

One way or another tanks ready for battle and fully equipped left a few days later for their intended purpose. It was these tanks in early November that took the brunt of the blows from the rebels rushing to the capital and helped defend Madrid.

After a few months, I had a chance to see some fellow tank men when they went on leave or treatment home. Among them were already Heroes of the Soviet Union and seeing them off in Valencia or Cartagena we remembered with pleasure the first steps on this previously unfamiliar and now so close land.

At the end of October, the Soviet transport "Kursk" arrived in Cartagena. On it were fighter planes, bombs, gasoline and a significant amount of flour in the holds. The flour was destined for Barcelona and military cargo should have been accepted here. The accumulation of transports in Cartagena made it difficult to immediately unload the "Kursk", as demanded by Prieto. Don Indalecio often called on the phone: "Don Nicholas, how is the unloading of important cargo going?" He asked and begged to speed it up not particularly worrying about secrecy when talking on a regular line.

The November battles for Madrid were approaching, and every hour was really expensive.

Despite the significant risk we decided to send cargo and transport to Alicante. This port was completely unsuitable for military purposes and poorly protected from the air. It did not meet the most basic safety requirements but what was to be done? Only the speed of unloading and the suddenness of such a risky event could ensure success.

A huge transport was moored at one of the berths and the population looked at it with curiosity but then was wary when huge boxes with "snub-nosed" I-15 fighters began to rise from the holds and behind them bombs. It might not have been very humane to endanger women and children, but the war claimed its preference and Madrid demanded help.

When the bulk of the cargo brought by "Kursk" was already on wheels or on the quays several rebel aircraft appeared over the port. There was nothing to repulse them with. The situation was extremely tense. An unexpected incident came to the rescue: at that moment, the Argentinean cruiser "25 de Mayo" was in the port of Alicante. The cruiser, in self-defense, opened fire on Franco aircraft. This worked and the rebels hastened to retire. Nevertheless, immediately after that we sent the "Kursk" to Barcelona and did not repeat such experiments in Alicante.

In November 1936, the government moved from Madrid to Valencia. There was heavy fighting for Madrid. After major changes on all ships relative order was established in the fleet. By government decree the committees were replaced by commissars and the squadron was controlled by the fleet commander. There was still heated debate in the city which at times turned into open clashes between socialists and anarchists, but the influence of the anarchists was clearly falling. The slogans of the Communist Party became the property of the masses and enjoyed popularity. Even those who were listed as socialists or anarchists often supported the communists.

November 1936 was difficult. The situation near Madrid at times became critical. The situation at sea worsened. The cruiser "Canarias", which surpassed all Republican ships in speed and armament, finally relocated to the Gibraltar region and entered the Mediterranean Sea. Republican transports coming from the Black Sea required substantial security. Cartagena was subjected to more powerful air attacks. The experience of unloading the first "players" revealed a lot of flaws. They worked slowly, the ammunition was handled carelessly and the anarchist unions continually disrupted work in the port dragging some of the workers with them. Fighters, tanks, guns, and rifles arriving in the first transports lay on the berths without adequate protection and were slowly taken away. It was no better on the railroad. Hundreds of wagons ready to be shipped due to lack of steam locomotives. All this became known to the rebels. Instead of the duties of a naval attaché I had to take on the work of the chief naval adviser. There is nothing for a sailor do to in Valencia and I kept a room at the Metropol Hotel only in case of my arrival. The main task for the coming months was to ensure maritime communications primarily with the Soviet Union. The difficulty of the work was that I could not order anyone and at the same time I had to influence the fleet commander, seeking the exact implementation of my advice. So on a strictly scheduled day and time fleet ships were to go to sea, meet transports with weapons somewhere off the coast of Algeria and deliver them to the port according to a strictly developed plan. I went to Valencia to I. Prieto. The Minister immediately accepted me. He had the habit of talking on the couch, his legs matched. Even frequent phone calls did not bother him; every time he got up went to the desk and then returned to his place again. So it was now. I talked about the expected operations of the fleet and asked him to give the necessary instructions to fleet commander M. Buiza and base commander A. Ruiz. Prieto seemed to be dozing but I already knew his manner of listening.

“I will give all the necessary instructions,” he said with a start and immediately called his secretary. He, as usual, appeared with an expanded notebook ready to shorthand all the orders of his minister. Don Indalecio dictated the necessary instructions.

“I ask you to personally report to me regarding all misunderstandings, which, however, I do not expect,” he finished and his eyes flashed from beneath his overhanging eyebrows. It was this instant that Prieto, whom the Spaniards pointedly called "Don Indus" raised his index finger. Outwardly clumsy with lazy movements, he was an old and experienced politician.

My personal participation in the campaign of the squadron to the North in this case was very useful. M. Buiza, whom I met more closely during the dispatch of the fleet to the North eagerly consulted with me on all matters and I tried not to infringe on his pride and wherever possible agreed with his recommendations. I trusted him alone with information about the time of the next transport approach in order to have time to prepare for the fleet's dispatch and he knew how not to divulge secrets.

In 1936, there was a special need for advisers in the base and on the destroyer flotilla. It was on these days that two of our volunteers S. Ramishvili and V. Drozd came to me. They appeared late in the evening on November 10 or 11, 1936. After hearing my native tongue I ran out to meet them and we joyfully greeted each other. The conversation dragged on far after midnight. Drozd and Ramishvili vied with each other how, trying not to give themselves away, they had not without adventure made their way from Paris, via Toulouse and Barcelona, to Valencia. Not wanting to call attention to themselves there, they fell into a hopeless situation, but then they ventured to call my nickname "Lepanto" and were sent to Cartagena without any difficulties. But the most pleasant thing was to hear their stories about affairs at home and especially about naval news. Listening to them I thought who should be left as adviser to the phlegmatic Antonio Ruiz and whom to send to the frantic Andalusian - the commander of the destroyer flotilla Vicente Ramirez. I knew these future advisers from duty at home and given their personality traits asked S. Ramishvili to assume the duties of adviser to the base commander and V. Drozd to work together with the destroyer flotilla commander. As subsequent events showed the choice was successful. Ramishvili who knew French quickly learned to speak Spanish, got along well with people and two months later knew the base with all its huge economy as well as don Antonio. From early morning he and his driver and friend José had already traveled along the berths of the port, talked with one or the other official and all this worked out wonderfully and the Spaniards, seeing his energy never refused to listen to his advice. Base commander Ruiz at first did not keep up with him and then calmed down having delegated many questions to don Juan Garcia (Ramishvili). All our volunteer sailors had to work on ships in difficult conditions. The socialists prevailed on some ships while on others ("Jaime I") the influence of the anarchists remained strong. Everyone wanted to participate in the struggle but often understood their role in different ways. Finally, not everyone treated our comrades alike although outwardly they tried not to show it. But the most difficult was the situation in the base. As in focus, all the contradictions were concentrated in the base and found expression not only in violent rallies or meetings but also in direct skirmishes with weapons in their hands. The anarchists whose influence in the province of Murcia was significant made a lot of noise but were the first to quit work when dusk set in and the likelihood of raids increased.

Don Juan Garcia had difficult responsibilities: to ensure the unloading of transports under conditions of frequent raids, and to his great merit in Cartagena we did not lose a single ship, nor a single rifle. In addition, he helped the Spanish comrades strengthen the anti-aircraft defense of the base, organize the protection of roads and harbors and the work of repair shops. Energetic, cheerful southerner, he somehow knew how to find a common language with the Spaniards and made them respect him. He knew people well and knew who to rely on. Around don Juan Garcia were collected together certain assets. So, on professional matters, he could rely on Colonel Mateo who honestly carried out all his instructions. He had his friends in the arsenal. While Patricio Casado, the head of the arsenal was not always loyal, his deputy Francisco Pereiro, secretary Melia and the head of the corps workshop Artajo were reliable assistants in any situation. It is worth mentioning "Almirante Lago" (as he was jokingly called), who carried out the heavy duty of the commander of the Vigilancia, that is, monitoring base's water area. Promoted by Juan from the junior officers, he hated the fascists with all the fibers of his soul and coped well with the growing economy, which consisted of different ships and boats. It was he who had to watch if any submarine made its way to the base during the bombing. The same loyal people were "in the Rocha Guard", which maintained order in the base, in coastal artillery and even in trade unions.

Valentin Petrovich Drozd by his nature would never have found a better use for himself than on destroyers. His main merit was the ability to influence his commander. Noisy and sometimes freaking out, Vicente Ramirez needed reasonable advice and perhaps no other adviser could do this more successfully than Drozd. If the cruisers did not always go out to meet the transport the destroyers must plow the sea between Algeria and Cartagena. In such circumstances Drozd and Ramirez quickly became friends and even in their free time were always together. V.P. Drozd repeatedly admired his behavior. That is how we knew Valentin Petrovich during the years of World War II. Death spared him in Tallinn and in besieged Leningrad, on the most dangerous campaigns during the evacuation of Hanko, or on the bridge of the cruiser "Kirov" during mass raids on this ship. And here is fate! Drozd died in 1943 when traveling by car from Kronstadt to Lisy Nos, failing into an ice hole created by a random bomb. "Savage death," he allegedly said when the car went under the ice. Escape was impossible.

Don Ramon, as Drozd Vicente liked to be called, was not good at knowing the language but was able to perfectly complement unknown words with gestures and soon did without an interpreter. The 15 destroyers in service required constant leadership. There were few experienced commanders and flotilla advisers were often forced to replace inadequately trained navigators, gunners, or miners. It was too much to train each one individually and commander training was organized in the flotilla earlier than the others. Ramirez met the first proposal about this with hostility. "Let's break the fascists, then we'll go to Cadiz to study," he laughed and Ramon convinced them without knowledge and the ability to use weapons, one cannot count on success. It was all the more difficult to convince the flotilla commander and ship commanders of the need for special ship sorties and formations at sea for torpedo and artillery fire practice. Yes, to tell the truth, we did not believe that the war would take such a protracted character and for a long time we counted on ending it in the coming months. Time passed. The squadron, along with the successful escort of transports, failed in the search and night attacks on enemy ships. "Burro, Cabrones!" Ramirez shouted but he nevertheless thought about the reasons for the unsuccessful actions of the ships. It became obvious that it was impossible to make plans on one's enthusiasm and then he turned to don Ramon for advice on how they act in this or that case in the Soviet Union. Valentin Petrovich patted his boss on the shoulder invited an interpreter and sat down to draw up a plan of classes or exercises. "Eh, if I could have put under arrest for 10 days he would know how to prepare for operations," Drozd once joke with me, but the trouble was that even in his position he couldn't even politely order. A personal example in battle and high knowledge were the only arguments.

By this time, the number of Soviet volunteers in Spain had increased. I would like to recall those such as V. Alafuzov, N. Basisty, I. Eliseev, V. Gavrilov, G. Zhukov, S. Soloukhin, I. Burmistrov, A. Alexandrov, who were in Spain when I was still in my life. Later in Spain arrived: L. Bekrenev, N. Egypt, N. Abramov, N. Pitersky, S. Sergeev and others.

I would like to talk about the future fate of some comrades in Spain.

V.A. Alafuzov worked at the Main Naval Headquarters for almost the entire war in fact being its chief. He played an important role in the operational training in the pre-war period and in improving the readiness of the fleets. In recent years, Vladimir Antonovich, headed the Naval Academy, mainly worked on theoretical issues of naval warfare. He died in 1966.

I.D. Eliseev during the war was the chief of staff of the Black Sea Fleet. He had to repel the first German onslaught in Sevastopol in October - November 1941. After the war with the rank of Vice Admiral, Eliseev worked for many years at the Main Naval Headquarters as the deputy chief of the General Marine General Staff. He is currently retired working on memories of his experiences. And he has something to remember. Like everyone who has been to Spain, Ivan Dmitrievich will never forget his first "baptism of fire".

N. E. Basisty in the republican fleet sailed on destroyers, returning to the Soviet Union he took command a cruiser and during the war years commanded a combination of ships on the Black Sea. Under the leadership of Nikolai Efremovich our ships broke through the blockade of Sevastopol and landed troops in Feodosia. After the war Admiral P.E. Basisty was appointed Commander of the Black Sea Fleet and later First Deputy Minister of the Navy. After retiring Basisty engages with former sailor volunteers in a veterans group.

A worthy representative of our country in Spain was G.V. Zhukov. I serve together with George Vasilyevich Zhukov on the

Black Sea in the twenties then fight in Spain and observe his work during the war years. It is no accident that in the most difficult months of the war, in the fall of 1941, by the decision of the General Headquarters he was appointed commander of the Odessa defensive area. Under his leadership until mid-October 1941, sailors and troops of the Primorsky Army valiantly defended Odessa and left only after the order of the High Command: "leave the city and evacuate to Crimea for the defense of Sevastopol." Hero of Odessa Zhukov remains in the memory of all Black Sea residents I recall that after the war, when he was sick with a paralyzed leg he asked me to give him all possible jobs. G.V. Zhukov liked to recall the "young years" and he always remembered the "Spanish" period of his life.

N.O. Abramov commanded the Danube Flotilla at the beginning of the war; N.P. Egipko - a combination of submarines. All these comrades tried to justify the confidence of the country. They shared the danger of military campaigns and like their Spanish comrades were under the bombs in Cartagena or on ship bridges.

I recall how in the winter months I happened to go to sea on the destroyer "Lepanto". Four Republican ships searched nightly for the rebel cruisers. The lead was "Lepanto" on which was also the flotilla commander Vicente Ramirez, promoted to captain de frigate of the Spanish fleet with his adviser. The destroyer was commanded by the famous Valentino Fuentes. He loved to brag about how, without "hesitation", he brought his ship to Cartagena to serve his homeland. However, by virtue of his character he could not take a firm line during the war. As the Republican situation became more difficult, he like many officers hesitated more and more until finally he slipped into the position of the monarchists. Fuentes was much older than Ramirez and considered him an upstart and himself not appreciated by the government as an honored officer.

- Hombre! - Suddenly it was heard on the bridge when the conversation between the commander of the ship Fuentes and the commander of the flotilla Ramirez turned to high tones and after that the whole bridge could hear something like this dialogue:

"Let me command the ship myself," Valentino said hotly.

"But your signalmen do not look well and so we will not see the enemy," shouted Vicente in a fasts breaking voice.

"I will complain to the Minister," was the response back and Fuentes began to get nervous.

- Que Pasa? (What happened?), Ramon intervened and calmly translated the conversation onto another topic.

One should not attach great importance to this. The two temperamental southerners quickly cooled. In addition, it got dark and we were in an area where the possibility of meeting with the rebel ships was ruled out.

Fuentes, as a hospitable host suggested that I go down to drink coffee, of course "with cognac" because on voyages this drink was always considered better than milk.

Leaving the bridge while sailing seemed unthinkable to me. I do not remember the case when I would leave the bridge even if the trip lasted several days. I preferred to stay in place now. They had not yet had time to bring us the "cognac" coffee when the signalmen reported lights observed "just ahead of the bow".

Vicente and Ramon bent over the map pondering their actions in case of meeting with the enemy. But the enemy unfortunately does not travel with lights and found the source of the lights soon outlined the silhouette of a small transport carelessly going southward.

The destroyers were moving at full speed in line ahead formation. Stormy white breakers from the stem and frothy waves behind the stern were well observed, showing the location of the ships. The moon had not yet ascended but a clear starry sky increased visibility and the horizon was clearly visible even with a simple eye. At the appointed hour the formation turned south to go through the area of likely "Canarias" traffic. When they reached the "main passage" several passenger and cargo ships crossed our course not suspecting our presence. We turned north again and having passed the allotted time lay down on a course to base. This was one of many unsuccessful searches. Only in March 1938 did the search succeed. The rebel flagship "Baleares" was discovered, torpedoed and sunk.

Ramon and Ramirez returned tired after such sorties and together, as inseparable friends, came to the base commander or to me.

"Salud and trimotores,"[40] Ramirez announced with his usual greeting, a joke.

"Don Ramon, muy listo", (that is, an intelligent man), Ramirez admired his adviser.

V.P. Drozd was actually a competent, brave sailor who did a great service to his Spanish friends.

To Valencia on the old road. Flight to Barcelona

It was at the end of March 1937.

"Salud, don Nicholas," I heard the voice of G. M. Stern. He called from Valencia. "I have an important message; can you be here in the morning?" He continued. It was already late but if I departed in an hour or two then I will have time to arrive in Valencia at dawn and come to him at Alboraya 8 where the headquarters of our main adviser is located. My driver Ricardo, who was dismissed home, will only appear in the morning and I asked S. Ramishvili to let his Jose go with me.

In less than an hour, my assistant A.P. Korobitsyn (Narciso) reported: "Coche listo" (the car is ready) and we went downstairs. After the first months of my stay in Cartagena I categorically refused protection and now my only constant companion was the translator - Narciso's assistant. A wonderful worker and a good comrade who lived in Argentina for a long time and knew Spanish very well, he replaced two anarchist marineros assigned to me by the base commander. There was a small Plymouth at the entrance to Capitania and Ramishvili gave the driver the final instructions.

"A sus ordernes" (at your service), - Jose loudly reported and stretched out military-style, reporting on the readiness of the machine.

"Vamos!" (Let's Go) - and the car drove along the familiar streets in complete darkness without turning on the headlights. Blackout was strictly observed during periods of frequent bombardment. There were few inhabitants in the houses at night and there were many homeless cats on the streets. Their green eyes sparkled in the dark which made some kind of unpleasant impression.

Jose was talkative. A native of Seville he loved his city. Whatever he talks about, at any opportunity the conversation transferred regarding Seville praising it in every way. So it happened now. As soon as the car drove away from the building of the base commander Jose began to compare the winding roads of Cartagena with straight and beautiful roads in Seville.

"Quen no ha visto Seville, no ha visto maravilla" (whoever has not seen Seville, has not seen miracles), reads the Spanish proverb. I remembered our Odessa. There a lot of jokes about the love of the local residents for her.

Having passed the outskirts of Santa Lucia with its narrow, crooked streets, the driver turned on the blue lights and we rushed along the familiar coastal road to Alicante. This road, usually empty even during the day was completely deserted at night. As a precaution, my Ricardo was even in the habit of taking a machine gun out of the trunk and placing it on the seat next to him when he passed through it at a late hour. He has been doing

---

[40] Literally "Cheers and Trimotors". Tongue in cheek reference to fascist tri-motor bombers which plagued everyday life in Cartagena.

this since two unidentified men stopped our car and asked to drive them to the nearest pueblo. Without suspecting anything bad I immediately agreed and it seemed to me that they were the most peaceful people. But Ricardo later assured me that it was a "pistoleros," that is, anarchists, from whom all sorts of nastiness could be expected. Since then the question of whether to take fellow travelers was decided by Ricardo himself and he used to step on the gas if someone raised his hand to stop the car. The night was bright and the lunar path stretched far across the sea. The rare village was empty and gloomy. Shutters are closed and only a passerby with a flashlight in his hand occasionally appears on the road and immediately disappears.

"It is a pity that you were not in our South," Jose continued his story turning back to make sure that we were not asleep and able to listen to him. He talked about how, before the rebellion, he traveled all the roads between Seville and Malaga, between Cadiz and Granada, when he was engaged in the transportation of fruits and wine, which is so rich in the southern provinces of Spain. He did not like gray dusty Cartagena and the gloomy roads in the mountains. "Montañas sip agua, mar sip pescado" (mountains without water and sea without fish), - he repeated the famous saying more than once. "And what of Granada?" Narciso asked him. José replied with a joke: "Quien no a visto Granada, no a visto nada", that is, he who did not see Granada, he did not see anything.

He remembered many sayings, sometimes surprisingly accurate, characterizing Spanish cities. He also knew the saying in verses about Cartagena. It said that there are mountains without greenery, sea without fish, sky without clouds. All drivers love to talk and Spanish ones doubly. And it would be nice if they didn't try to combine fun chatter with fast (in Spanish) driving and didn't get into the ditches so often.

So, we drove quietly into Alicante and turning off the headlights, slowed down. Jose asked permission to stop near the hotel to inspect the car before the long stretch to Valencia and we were asked to go for a coffee and warm up. My legs were really tired in the small car, but I still could not get rid of the consequences of a car accident and after a long trip I could not straighten my left leg right away - my knee was badly swollen. Unlike Cartagena the city lived a calm, peaceful lif, and in the roadstead there was always some kind of military or merchant ship fully illuminated. So it was now. In any case standing away from the port a small ship swayed; judging by the silhouette it was a military destroyer busy controlling incoming cargo according to the instructions of the non-intervention committee.

When leaving Alicante, to our left, there was a dark silhouette of an old fortress, almost as old as Cartagena. I was falling asleep, and besides, Jose, dissatisfied with something, was silent.

The road was familiar. We already happened to ride on it many times and knew all the major cities or pueblo. This alerted us. At least half an hour passed and we did not meet a single settlement. Small rubble and pebbles crunched under the wheels which also seemed suspicious.

"Are you on the road?" I asked Jose.

"Segurisimo," he replied, "there is no doubt." It seemed uncomfortable to teach the driver but I still asked if he had ever traveled to Valencia. "Many times, however, a long time ago," he replied. We continue on the path. An extremely flat road made of small pebbles and not a single oncoming car. Something was amiss I thought but decided to remain silent for now.

After driving for another half an hour, I again asked Jose if he was sure of the correctness of his chosen path.

"Seguro" (I'm sure), but without a superlative pitech, he answered and explained that this road is much shorter than the one that is usually used now. "But why doesn't anyone use it if it is shorter?" Narciso asked him. "This is a little more dangerous because the road goes to the mountains where there are almost no villages," José explained gesturing that we had nothing to fear, we were armed. Again we drive on and again not a single oncoming car or at least some village. If I weren't sure of Jose's honesty and loyalty, I would immediately order him to return and then there's a false shame for excessive fear so we continued on into the unknown. But finally a pueblo. We happily stopped and sent Jose to find out where we are. Jose goes from one house to another, knocks on windows - and all in vain. The has village died out. Outwardly everything looked normal and seemed to speak for people living in it but why didn't they open the doors? Mysteriously! On the outskirts of the village we find a road sign with an arrow and the inscription: "Alcoy". This is good at least there is some kind of landmark.

An army comrade would definitely have a detailed map of the area with him and he would easily figure out the roads and distances.

True, there is a well-known joke that when two army officers got out of a car at some intersection and began to look around, unfolding a map, the boys nearby noticed: look Petka they will ask for directions now.

We, the sailors, trusting the driver on the familiar road, had and even then by chance, a simple geographical map for tourists. It was clear from it that two roads go from Alicante to Valencia: the older and once, apparently the only one, went along the direct path, and the other built by Primo de Rivera and meeting the requirements of tourists, passed closer to the sea along tourist centers, restaurants and resorts built there. Our Jose knew only the old road well and did not dare to go at night on the new one. But the war brought about changes. One road turned out to be completely abandoned and apparently not because of the danger of driving in the desert but because of the lack of proper maintenance there and above all lack of gas stations and the other took on the whole burden of communications. Having calculated that it is now closer to Alcoy than to return to Alicante we decided to continue on. After a smooth and relatively straight path, the road suddenly began to climb uphill. The car moved slowly and described large circles around the mountain climbing the pass. Low clouds covered the top of the mountain, and we drove in dense fog, reducing speed. We were still mystified regarding why we had not yet met a single car.

"And what about gasoline?" I asked, thinking that if, God forbid, we are left without fuel or the engine turns sour, then we will be sitting there without counting on help.

"Bastante" (enough), - Jose answered and looked at the dial, but there was no confidence in his tone anymore, and he began apologetically to explain that when we went up the pass, then we would go on a continuous descent all the way down to Alcoy. We really soon began to descend in the same circles and ten minutes later we left the clouds. It was already past midnight and instead of taking a nap, for which I always used night trips, we were all awake carefully and cautiously peering into the darkness. One consolation is that now we were going down and the fuel consumption was low.

Suddenly, we saw blue lights and silhouettes of houses - it was a city, the same Alcoy, which we needed like air. There was almost no gas in the tanks.

It is difficult to say who was most happy about this but the driver was clearly pleased. Having approached the control post he jumped out of the car and animatedly began to share his feelings when the "miliciano" demanded from him "salvoconducto" (pass), looking suspiciously at the car.

The traditional salud, and we moved to the column to refuel with gasoline.

"Now everyone prefers to ride along the coastal road," José said and added that the miliciano was also surprised by our choice of road.

"This is not Seville for you," Narciso joked when the car was already racing along the paved highway to Gandia from where it was not far from Valencia.

It was lively at the Metropol Hotel in late March 1937. Everyone lived through the Republican victory at Guadalajara. There was a lull after the violent attacks of the Italian expeditionary force and its complete defeat. This was felt in Valencia. In the evening the hotel restaurant is unusually crowded. Journalists and writers from England, France, the United States, who visited the battlefield now gathered in Valencia. They were interested in further plans for military operations and a looming government crisis. Many sought to visit the hotel where our office is located and therefore you can meet there not only representatives of the Soviet and Spanish press, but also the Frenchman Malraux, the American Hemingway and many less prominent figures. Spring is in full swing in Spain in March, not our April or even May in Moscow with cold nights and often frosts, but hot. It is already quite warm during the day and if it rains it is warm and short, a real spring rain that fills rice fields and orange groves with moisture. After a sleepless night I managed to relax a bit waiting for a call from Stern.

"Something is amiss in Barcelona," said Grigory Mikhailovich immediately after greeting him when I came to him. As always In his room on the second floor, cards are laid out on the table. There are no all-seeing journalists and the situation with its severity resembles the headquarters of a large operational compound.

"Prieto is going to send several ships there," he continued, wanting to find out my opinion on this subject and give some instructions. I have already heard from Commander Buiza that the cruiser "Mendez Nunez" and two destroyers are preparing to leave for Barcelona, ostensibly to provide escort for important transports from France to Catalonia and Valencia although the departure time has not yet been specified. Having told Grigory Mikhailovich about this I asked to inform me about the situation in Barcelona.

"The anarchists put forward some demands to the Caballero government and the case could end in an open clash," Stern said, perceiving this as a common occurrence. Relations with Catalonia have indeed exacerbated more than once and negatively affected the course of hostilities. Much depended on her in the fight against the fascists. "Catalonia is the vital nerve of the Spanish Republic," D. Ibarruri once said. In Barcelona, the cotton, chemical, and electrical industries were concentrated. The outcome of the war largely depended on the ability to mobilize the industrial and human resources of Catalonia. However, instead of consolidating efforts to defeat Franco, the anarchists demanded "complete independence" or the adoption of their military plans, in particular on the Aragonese front.

As you know anarchism in Spain had a rather solid base especially in Barcelona - the birthplace of world anarchism. The desire for independence and the unwillingness to obey the central government prevented the use of the huge potential opportunities of industrially developed Catalonia. This was especially scary at a time when common efforts were required to fight the enemy. Meanwhile, it is known that if a Catalan goes from Barcelona to Madrid or Valencia, he says: "I am going to Spain." If you ask him: "Isn't Barcelona Spain?" - he will answer that it is "a habit and no more", but he will persistently continue to say: "I am going to Spain", emphasizing that Catalonia is something else. Knowing all this I was also not surprised at the next disagreement between Valencia and Barcelona.

I promised to find out from Prieto his intentions in sending ships. And we turned to the question that is always topical for Stern: when will the next "players" arrive, how does the fleet intend to ensure its arrival and will there be a delay in unloading the arriving aircraft, tanks and guns.

In the afternoon, I called the Secretary of the Minister and asked to receive me. Prieto, who usually scheduled the hour of the visit without delay, said that he apologized and asked me to arrive the next morning.

At the appointed hour, I was already in the reception of the Minister.

Prieto immediately accepted me, but, not listening, suggested that I go with him immediately to Barcelona.

"The plane is ready and in half an hour I have to leave," the minister said, collecting some papers in a briefcase. "We will solve all the issues along the way," he continued, and I had no choice but to answer: "si, señor"

I asked for permission to call in to the Metropol to report on my departure and I promised to arrive at the airport in a timely manner. Calling Cartagena, I found out that the ship formations had already received the order of the Minister and was urgently leaving for Barcelona with unknown goals for the time being. "Under these conditions I would rather fly there myself to follow Stern's instructions," I reasoned. Prieto's proposal therefore completely suited me and sharing my thoughts with Ramishvili I hurried to the airport.

A tall and dry general was already pacing around the plane, it was Ignacio Cisneros - Air Force commander who arrived to accompany his minister. Cisneros following Prieto was supposed to fly to Madrid where Republican aircraft continued to operate actively in the Guadalajara region. "Barcelona buena" (Barcelona is good), said don Ignacio and advised me to find time to see this city with its beautiful streets, buildings of amazing architecture and the most beautiful metro in Spain. I remembered the comic dialogue that Barcelonaers supposedly often exchange with residents of other provinces.

"Barcelona buena" (Barcelona is good), - residents of Barcelona like to brag.

"Barcelona buena si la bolsa suena" (Barcelona is good if your wallet rings in your pocket), they object.

"No tanto si suena, como si no suena, Barcelona siempre buena", the Barcelona responded, saying that regardless of whether the wallet rings or not, Barcelona is still beautiful. These phrases are sometimes cited as an example of the sonority of the Spanish language.

But Prieto's car appeared and the crew of the aircraft began to fuss making final preparations. The short and fat Don Indus was surprisingly mobile when necessary. He energetically jumped out of the car and literally on the move, giving some instructions to Air Force commander General Ignacio Cisneros and headed for the plane. The sun was pretty hot. Immediately after takeoff the whole Gulf of Valencia is spread out under the wings of an airplane. The famous orange groves were an amazing sight. Clinging to the coast at a low altitude we flew to Barcelona. To the left in the distance, the old fortress of Sagunto was barely visible. In its place, the new city of Sagunto arose - the center of the steel industry. Once Hannibal who had been at war with Rome for many months attacked this fortress and after taking it after a long siege was forced to cede it to the Romans again. Like many cities in Spain Sagunto is associated with the centuries-old history of the Iberian Peninsula.

To our right beyond the horizon were the islands of Mallorca and Ibiza, occupied by the rebels. They began to play a more significant role later and at the time described only rebel ships were temporarily based on Palma and planes bombed the Republican coast in small groups from there. The sea was calm, only small ripples were visible and a white strip of surf stretched along the coast. It is incomparably more beautiful than in the south of Spain with greenery scorched by the sun, gloomy rocky shores and gray gloomy buildings of Cartagena.

I sat opposite Prieto and waited until the minister having gotten comfortable with the situation in the air began a professional conversation. He was preoccupied. His mission was delicate. He was supposed to help President M. Azana who was in Barcelona conduct important political negotiations and calm the raging anarchists. As an experienced politician Don Indus was actually perhaps better than anyone else able to accomplish this task. “A strange situation,” I thought, looking at him, “a man who had never known aviation or the navy before, now at the time of his country's trials has become the head of two military ministries and the correct conduct of warfare on the sea and in the air depends on him. But such is the structure of many capitalist states. Reasonable politicians without a special military education but knowing how to rely on good advisers are able to lead military ministries and direct the activities of the military. Such people are not burdened by traditions and knowledge of only one specialty; they can objectively weigh the pros and cons. They should be able not only to subordinate strategy to politics but also to try policies to the capabilities of their ministries. Experience shows that a politician who is called to lead a country or responsible ministries, engaging in politics and determining the strategy that follows from it, must carefully delve into the essence of this strategy and persevere in practical matters. Otherwise, the policy will be without the foundation on which it is built. However, I moved away from the topic. Prieto was too much a politician. He thought a lot about his political successes and did not do enough about the affairs of the ministries entrusted to him. So, without hesitation he sent the entire republican fleet to the North in September 1936 more concerned about his prestige in the next Cortes elections than about the military expediency of this maneuver. Now without consulting the fleet commander M. Buiza he sent warships to Barcelona to reinforce his political actions there during the negotiations.

Placing himself more comfortably he carefully and slightly ironically looked at me.

"Ola, Nicholas," he began as always a conversation with me.

Asked if I was aware of his order to send ships to Barcelona, Prieto did not say that they might be required in the event of a critical situation in the city, and explained the presence of the fleet in Barcelona only as a courtesy visit and a sign of friendship and unity of purpose in the fight against the Nazis. He apparently did not want to be very frank with me, a foreigner, and I did not consider it convenient to insist. It was not difficult for me to find out the details of the situation without Prieto's help. I asked the minister how long the ships would be busy in Barcelona and whether they would return to Cartagena for the intended sortie to protect communications. Prieto said the safety of transport arriving from the Soviet Union remains the most important task of the republican fleet and he will take all measures to quickly return the ships to Cartagena.

"When is the next transport expected and what are your complaints about the fleet," don Indalecio asked moving the conversation to another topic. He was pleased that so far no ship had been lost and that all operations to unload them under bombing had been successful. I in turn praised the actions of the Spanish sailors and asked if the fleet commander complains that sometimes I inform him late about the traffic. "Keeping secrecy is the most important thing for the success of the operation," I assured him of anticipating such claims. "It would be nice if you yourself came to Cartagena," I said, but Prieto did not react at all to my proposal. The Minister clearly did not attach importance to the actions of the fleet although he understood that the formation of army units depended entirely on airplanes, tanks and other equipment coming in by the sea.

It was not until June, when the battleship "Jaime I" exploded, that he went to the main fleet base – for political reasons, he could not help doing so.

The navigator came in and laid out a flight map in front of the Minister. We flew over Cape Tortosa and Barcelona was less than half the way. A salient in the Zaragoza region marked the front line. Here the anarchists have repeatedly vowed to "win or die," but the matter did not move beyond the loud slogans. Zaragoza remained in the hands of the rebels. Prieto shook his head and with a gesture released the navigator.

Taking the opportunity I decided to tell Prieto about some of the naval problems. The weakness of the aircraft attached to the fleet was felt more and more every day. Without having a formation of medium-sized bombers and almost as many fighters it was impossible to reliably cover the fleet's base anchorage and its operations at sea. "The fleet needs more planes," I said, apparently unexpectedly for the minister.

"Si, Si," Prieto repeated several times, but declined to give a concrete answer to my request. Instead he started a more abstract conversation about building aircraft in Spain adding that he would talk about this in Barcelona. "I'll have to talk with General Douglas (Smushkevich)," I thought and turned to the naval theme.

We discussed how to speed up the repair of the cruiser "Cervantes", which was hit by a torpedo[41], what operations the rebel fleet can under tale, it was already based at that time not only on Straits of Gibraltar but also on the island of Mallorca; we talked about how ships need to engage in combat training and even touched on the issue of fuel reserves in Cartagena. But all these conversations were general in nature and did not take the form of specific decisions. The Minister was already mentally in Catalonia and was delighted when I, leaving him alone, began a conversation with his secretary about the beauties of the northern coast of Spain and the antiquity of the Barcelona port.

When we flew up to Barcelona, Prieto, as a polite host, asked where I intend to stay kindly offering to live in the same hotel where the room was reserved for him and immediately ordered it. In fact it turned out differently. The minister was visiting President Azana and my room was also empty: I preferred to use the room at our consulate with V. A. Antonov-Ovseenko.

After parting with Prieto at the airport, I did not see him again. He spent days on ships or on city tours and in the evenings everyone gathered at the consulate.

With the commander of the cruiser "Mendes Nunez" - Prado, we talked a lot about the situation in Catalonia and the anarchists. Fortunately the situation quickly changed for the better and after a few days the ships went to sea, heading home to

---

[41] As mentioned on page 52, an Italian submarine torpedoed the cruiser.

Cartagena and I went by rail to Valencia to report to my superiors about future fleet operations.

War continues

Spring 1937 arrived

The Republican army with the help of the Soviet Union and volunteers from international brigades stopped the fascist attack on Madrid, defeated the rebels near Guadalajara and began preparations for the attack.

The government has changed for the better. Largo Caballero had not lived up to the expectations of the people, left; he was replaced by Juan Negrin. There were new hopes for a successful end to the struggle. The supply of weapons was crucial. Germany and Italy sent the rebels whole formations of aviation, tanks and infantry. The Italian ships, without even disguising themselves, acted against the Republicans. To replenish the rebel squadron Franco were "sold" new destroyers and submarines.

Having not succeeded on the Central Front the rebels, as you know, in February 1937 launched an offensive in the south and captured Malaga. Now they were preparing for an attack on Bilbao.

By this time significant changes had occurred in the republican fleet and in Cartagena. The squadron under the command of M. Buiza began to understand that, in addition to enthusiasm, knowledge and daily practice are required. Ships went to sea for firing exercises and for joint maneuvering although this was rarely done and not very willingly.

Several dozen volunteer sailors from the Soviet Union sailed on various ships, helping young officers and personnel. The fleet was given a certain number of bomber aircraft and fighters covered the base in case of raids. Life in the city has also changed. After the winter bombing the population learned to take refuge in shelters and continue working immediately after the end of the alarm.

All attention was focused on the work of the port and ship repair workshops of the arsenal. Before the next transport arrived the comrades concerned began to worry. Most often I had to deal with pilots and their leader, chief adviser on aviation, Y. V. Smushkevich. Smushkevich liked the name of the Spanish battleship “Jaime I”, and he made it a household name for any ship. Jokingly he did not call it “Jaime I”, but simply “Haim”. If he wanted to find out about a transport he was interested in, he would ask: "How is it with you, has the next Haim has arrived?" And if I asked him for fighter covers of some warship, he answered: "Oh, these are for your Haimas," but always helped. For more detailed negotiations, Yakov Vladimirovich Smushkevich (General Douglas) usually came by himself.

Tanks appeared less often in Cartagena but if a large batch of tanks was expected then General D. G. Pavlov the senior tank volunteer called. Talking with him on the phone was dangerous; he did not recognize secrets and would directly ask when the transport was expected and how many tanks were on it. He did not even consider it necessary to call them “turtles,” - this offended him. It always seemed to me that Dmitry Grigorievich lacked the organization and personal fit that are so necessary in military affairs.

In Spain I did not happen to meet often with him but in Moscow I saw him many times. “Combat commander,” I thought of him, but I was still surprised to learn of his appointment to the high post of combined arms commander. He "burned out" in the first weeks of the war and only historians can tell to what extent he was guilty of losses and retreat in these difficult days.

Nikolai Nikolaevich Voronov was interested in artillery. “Well, has the groom not yet met his bride?” Asked Voltaire (N. N. Voronov), believing that such allegory does not reveal secrets. “Yes, everything is in order, we met and arrived in Cartagena," Voltaire had to be deceived in order to simultaneously mislead the enemy.

Regarding Nikolai Nikolaevich Voronov, I most often saw in Valencia. There on Alboraya street 8, was our main adviser - first J. Berzin, and then G. Stern. Voronov's tall figure stood out among others. Dressed in civilian clothes, in a black beret, he never lost military bearing. He was always cheerful, happy and loved to joke. "Scott has already come out of me, but Walter is not there yet," he laughed when I called him alter Scott instead of the official Voltaire. In fact, NN Voronov was called in Spain Voltaire - Colonel Voltaire, as indicated in his document. So not the English writer but the French philosopher was his name sake. One way or another, but Nikolai Nikolaevich was a real Russian in appearance and character. Last name Voltaire badly disguised him.

He also stopped in Cartagena when there were a fair amount of artillery and shells on the transports. In Albacete where new brigades were being formed weapons were always eagerly awaited. N. N. Voronov was "rooting" for his job, helping me send the equipment as intended.

Fate brought us back to Moscow. All the years of the war, I met with Voronov at Headquarters or at the People's Commissariat of Defense. His activities are known and do not require additions.[42]

Once in the reception room of the Soviet ambassador at the Metropol Hotel in Valencia I met General Petrovich. After the November battles for Madrid he arrived in Valencia for information. Invited to a cup of coffee to the ambassador, we got into a conversation and, as sometimes happens, were imbued with mutual sympathy at first sight. Many times I met with Petrovich on Alboraya Street, 8, where our advisers were located during the raids in Valencia, we went there but I did not know his real name. There was no need for that. He simply called me Nicholas, and he was for me companero Petrovich. Then General

---

42

Petrovich left for his homeland and I quite by chance met him on Gogolevsky Boulevard on the day I arrived in Moscow at the end of August 1937. Ola! - exclaimed not General Petrovich, but Kirill Afanasevich Meretskov dressed in Soviet uniforms. Since then, less often, then more often we met with him before the war in the General Staff, during the war years on the Karelian Front or in the Far East.

I did not meet Colonel Malino in Spain and only heard about him several times in connection with the battle at Jarama. Of course, it did not occur to me to simply decipher Malino as Malinovsky Rodion Yakovlevich.

I only found out about this in 1948. Having not seen him more than once during the war years I did not suspect that the corps commander near Odessa and then the army commander near Stalingrad and finally a front commander at the end of the war, Marshal R. Ya. Malinovsky was an adviser to Republican army at Jarama during the years of the struggle against fascism in Spain.

But in 1948 our service paths converged. He commanded troops in the Far East and I, by the will of fate, turned out to be his deputy for naval aspects. I had to not only work in Khabarovsk, but also make joint trips on ships. The land is vast - from Kamchatka to Port Arthur. If the weather was calm then there was enough time to talk about current affairs and to recall the Great Patriotic War.

So once former Colonel Malino started a conversation about how he had to spend many months in Spain which he will never forget. Memories became common. Madrid, Valencia, Barcelona, 5th regiment, inter-brigade, part of the anarchists, republican fleet. General acquaintances: V. A. Gorev, K. A. Meretskov, N. N. Voronov, G. M. Stern, Mate Zalka.

He listened with interest about the republican squadron which he knew very little about before and I got acquainted with many interesting details of ground battles. It turns out Malinovsky was an adviser to Lister when the first divisions were just forming in the Spanish national army.

Later reading the memoirs of Malinovsky I imagined everything so vividly as if we had met more than once there, on the Iberian Peninsula.

When a transport arrived in Cartagena not only the Spanish stevedores took part in its unloading but also pilots, tankers, and volunteer artillerymen. They were interested in the earliest possible receipt of equipment and weapons and prepared transportation means in advance. Immediately on the spot in Cartagena destination points were indicated.

Aircraft as a rule were transported by truck to neighboring airfields and assembled there for testing and dispatch to the front; tanks were loaded and went by rail to Archena, Albacete and Madrid, and guns and ammunition were sent to the places of unit formations of the people's army. On the days of the arrival of transports berths usually experienced a great deal of activity. Spanish sailors knew many Soviet comrades and met like old friends. In the evening in a small club, or rather adapted for this, an apartment of five rooms, a friendly dinner was arranged (sometimes with smoked sausage and vodka sent from the USSR) and a gramophone was started. The record with the song "Seven Brave" was very popular and was listened to several times in a row. The radio tuned to Moscow and all thoughts involuntarily transferred to the homeland. Unfortunately very often such days were accompanied by bombings that interfered with our plans. This was the case for example on May 1, 1937 when the large Spanish transport “Santo Tomas” with valuable cargo arrived in Cartagena. Early in the morning he moored at the main pier of the port. Cartagena has already gained extensive experience in organizing unloading with the least loss of time. They knew that “guests” - “heinkels” or “fiats” - would certainly appear at night - and transport might have to be taken out of the port. The Spanish comrades in charge of receiving goods had gathered the day before with the base commander Antonio Ruiz and discussed how to get down to work faster when the transport approached the berth, how to prevent the accumulation of dangerous cargo in the port or at the railway station. In this case the difficulty was that on this transport arrived large-sized crates with bomber aircraft, tanks and even torpedo boats. Ignacio Cisneros, the commander of republican aviation, also flew to Cartagena. He was worried not only about transporting bulky fuselages of SB aircraft to nearby Los Alcazares or San Javier airfields, where they were going to, but also about guarding the entire area from possible rebel raids. If the enemy has still not been able to sniff out about the movement of transport, which is unlikely, then with his arrival at the port there could be no question of the secrecy of the operation. The whole city of course found out about this as soon as the huge black hull of the liner "Santo Tomas" with the help of tugs began to be pulled into the harbor. Only the barrage of fighters and the complete readiness of anti-aircraft artillery could now guarantee successful operation in the port. With the "player" at the pier and fighter planes, albeit in small numbers but constantly in the air, everyone worked calmly and confidently. Shore cranes and ship winches were lifting heavy loads from the upper deck and port workers stood ready in the holds waiting for their turn to proceed with unloading. Railway platforms and lorries with trailers were concentrated at the transport's rails. In a word the work began to boil.

As usual, Soviet volunteer pilots from neighboring airfields, Los Alcazares and San Javier arrived in Cartagena to meet their cargoes and comrades. On May 1, General Douglas came to Cartagena, he was already a hero of Guadalajara and was known by many Spanish comrades. When he once appeared with me on the cruiser "Libertad", he was quite rightly greeted by exclamations: "Viva, Russia, Viva!" A lot of planes arrived on the liner from this voyage. Huge boxes with disassembled bombers were to be unloaded as soon as possible and sent by special vehicles to the assembly place. Dispersed over a larger area they were less at risk than in Cartagena. In addition to aircraft on the upper deck, under covers, there were torpedo boats. They were intended for the republican fleet. Volunteers arrived using a passing flight. The ship had not yet managed to cast off moorings and exclamations of salutations between shore and the ship were already carried in Spanish and Russian.

“Salud, Amigo!” (Hello, friend) - someone shouted to the captain of the transport, whom many knew.

“Hello, Grisha!” - was heard in Russian. It was the sailors who recognized their comrades.

Now, along with the republican sailors several dozen of our volunteer sailors fought. The chief of staff of the base and adviser S. S. Ramishvili, like Figaro, appeared here and there in his little Plymouth, giving the necessary instructions in Spanish which he really mastered "Perfectamente Como Cervantes" (Perfect like Cervantes).

On the cruiser "Libertad" with the fleet commander worked (before the arrival of V. A. Alafuzov) N. P. Anin, who arrived in Spain in September 1936 and managed to participate in the northern campaign. I remember how I first met him at the Palace Hotel in Madrid. He had a large full beard. To the question of what to do, I had to answer: "Immediately shave your beard." I don’t know how he survived this but after half an hour or an hour he came to me and cheerfully reported that the order was fulfilled.

For almost half a year he sailed with the commander of the destroyer fleet, V.P. Drozd, whom even Russian comrades always called don Ramon, he enjoyed the great love of those around him.

I. Burmistrov served on submarines and did everything possible for successful combat operations of submarines, fulfilling commander duties of a type “C” boat.

All of them, and many others who arrived earlier or later sincerely sought to help our Spanish friends and shared with them the dangers and hardships of maritime service.

A group of volunteer sailors also arrived at “Santo Tomas”.

On the upper deck I managed to track down my old comrade V. A. Alafuzov who was entrusted with delivering this "cargo". He certainly survived many troubled minutes when he was met by the republican squadron and safely entered port. When the ramp was lowered and base commander A. Ruiz climbed the bridge warm greetings were exchanged on the deck and on the pier. A meeting in a foreign country even with an unfamiliar Soviet person can be extremely pleasant. This was doubly felt then in the military situation and the hugs of those who met were sincere. People met as close friends although many saw each other for the first time.

Despite the nightly alarms the berths that day were lively, fun and even a bit solemn. May Day holiday left a definite imprint on everything. No matter where the Soviet man was he remembered how solemnly and cheerfully this day was celebrated in his homeland. The desire to feel together with his people manifested not only on such days. For example we often listened to the music of the Kremlin chimes at midnight and when in the middle of this motif we heard the sound of a car horn and it seemed that we were seeing cars rushing along Red Square. It was especially nice.

In Cartagena, the first stage of the delivery of military goods to Spain ended. Then the second stage began: everything received had to be urgently unloaded and sent to various points of assembly, testing and unit formation.

Airplanes were sent to neighboring airfields. There, scattered throughout the airfield and even along the highway, they were put into readiness. At airfields it was sometimes no less hectic than in the city. When there were no ships in Cartagena rebel raids transferred to airfields where many of our comrades who assembled aircraft lived. In their free time they liked to come to Cartagena to drink a cup of coffee or a glass of cognac while they preferred that a cup of coffee be smaller and a glass of cognac larger.

We had a military friendship with the Spanish pilots and our comrades who worked at the airfields. Their fighters covered the fleet and when unloading occurred they patrolled over the base. This guaranteed us against daytime raids. Bombers operated at sea with ships or bombed the ports and ships of the rebels. A small number of them were given to the fleet and if necessary we asked General Douglas to send us a couple of squadrons which he usually did not refuse if the situation on the fronts allowed.

This formation of Katyushas attached to the fleet was headed by I. I. Proskurov and N. A. Ostryakov. They were especially closely associated with the republican fleet. After unsuccessfully bombing enemy cruisers both came to the ships stationed at the base and getting to know the nature of their targets developed tactics for future operations. They were wonderful people and brave pilots. In all weathers and over long distances they went to sea as a unit or in separate planes and were terribly upset when the bombs they dropped seemed to hit the target and the cruiser still went away without damage.

Often there were tanks in Cartagena. Having received tanks in the port they roared around the city with a roar and often fought on the move somewhere near Madrid. As a rule these were Soviet volunteer tankers, who after hot fighting, considered the days spent in Cartagena during the unloading of the "players" to be rest.

All volunteers - pilots, tankers or sailors did their job with honor. Having received combat names and surnames, they passed or crossed the Spanish border and soon dressed in "mono", with berets dashingly pulled over the back of their heads, already mastered the necessary Spanish words.

I remember once returning from Barcelona to Valencia by rail, I watched several young men get into the car accompanied by a girl-translator. She gave them final instructions. “They will meet you in Valencia but for now settle down here,” she pointed to the compartment of the carriage. Young travelers who had just arrived through France kept together, talking quietly among themselves. I wanted to talk with them but it did not occur here.

"Where are you going?" I asked a routine question to begin with. Glancing at each other, the two young men decided not to divulge their plans to me, the elder answered clearly: "Yes, we do and we are going" - and let me know that he didn't want to talk about this topic any further.

Knowing where they bound for Valencia I decided to make the best of it and half an hour later offered them my services adding that I do not want to ask them anything anymore. The conversation gradually revived. When the comrades who met them at the train station in Valencia explained who I was they apologized embarrassedly.

Of all the transports that made several dangerous voyage from Cartagena to the Black Sea, the most brave were the "Santo Tomas" and "Magallanes" mentioned above. The first one at the end of 1937 after I left Spain was sunk by its crew during an attack by the rebel ships and the second ceased its voyages at about the same time due to the then established superiority rebel forces at sea and open intervention of the Italian navy .

There are two events connected with the last voyage of the "Magallanes" that should be mentioned, but before that in keeping with the chronology, a few words about the mining of the British destroyer "Hunter".

English destroyer Hunter blown up by mine

In May, an unpleasant incident occurred in Almeria. Spanish sailors were already accustomed to English destroyers observing the roads of Cartagena and Almeria, formally fulfilling the role of observers of the "non-intervention" committee and often engaged in petty smuggling. Unlike German ships they did not conduct open hostile activities and did not inform Franco's fleet about the movements of the Republicans. The attitude to the English ships was friendly. They did not approach close to the coast and in the case of bombings they retreated even further into the sea. Brightly lit in this case they increased speed and plying out of danger waited for the artillery cannonade to subside and the bomb explosions to stop. Without risking anything they watched the base fend off the violent attacks of fascist aircraft and how houses continued to burn for a long time after the raid and sometimes the ammunition and fuel depots exploded. This usually happened at night. At dawn the English ships came closer to the port or base, anchored and on hot summer days they even put awnings for coolness. The only inconvenience for the English sailors was the inability to go ashore to play tennis and have fun somewhere in a restaurant. In general, this was a rare case when the war is not like a war contrary to the French proverb.

And so several days after the Almeria raid the English destroyer "Hunter" was standing in. The city and the port were often bombed and the destroyer used to anchor at a certain distance from the port. On May 13 at 2 p.m. a deafening explosion was heard. All those on the streets looked around with concern, expecting another disaster. But everything was calm - no alarm, no firing of anti-aircraft guns or machine guns. The enemy was not visible from the sea either. They did not immediately notice that there was an explosion of a mine accidentally floating off the side of a neutral warship. The mine does not understand - who is his own and who is "neutral". It is possible that before that she had been floating for a long time in a half-flooded state, barely noticeable even upon careful observation or maybe completely immersed in water. But in her way was the hull of a warship. It strikes and then everything happens as it should: a glass ampoule breaks and some kind of "Green's liquid" acts on the detonator which in turn causes a strong explosion of tetronitromethylaniline or TNT as it is called to simplify it. Maybe everything was different but with the same results. The force of the explosion was so great that no one aboard can withstand it much less the hull of an unarmored destroyer. Water begins to flow into the huge hole and the fate of the ship is unfortunate. The "Hunter" listed quickly and his position seemed hopeless. Apparently the captain assessed the situation as such and ordered the crew to abandon the ship. After removing the wounded the boats departed from the ship which with a list of 25 degrees continued to stay afloat. The emergency team of the port of Almeria quickly helped the abandoned ship and according to international rules could even claim it as their property.

The English sailors returned to the ship not without embarrassment over their hasty evacuation.

The next day the cruiser "Arethusa", taking the destroyer in tow, took him to Gibraltar. 8 men were killed and 14 wounded as a result of the explosion. The Spanish sailors regretted the death of the British though they did not help the fighting Republicans but also did not help the rebels.

Incident with the German battleship "Deutschland"

By May 1937, the struggle for every transport carrying military cargo from the East required the participation of the entire fleet squadron. The rebel cruisers based on Mallorca occupied a flanking position. German and Italian warships rendered them all kinds of help. During one such operation a sensational incident occurred in late May with Spanish aircraft bombing the German battleship "Deutschland" and the subsequent bombardment of the city of Almeria by the German military vessels.

Events unfolded as follows. On May 31, a squadron of republican ships came out to meet the next "player" - in this case, the "Magallanes". Information about the appearance of rebel cruisers forced the fleet command to mask their intentions in order to divert the enemy's attention from the intended meeting point. Having selected the port on Ibis Island as the object of diversion, the republican fleet went there to shell the port and after dark proceed to the meeting place with the "player" near Cape Bon. The transport was large and the cargo on it required special protection.

Going to the port and finding there a German battleship the fleet commander M. Buiza decided not to complicate the international situation and refused to shell the coast. Moreover the main goal - to attract the attention of the enemy - has already been achieved. The squadron turned south. But in addition to ships aircraft also participated in this operation.

At the appointed hour an aircraft formation appeared over the island for a simultaneous attack with the ships. However it is much more difficult from aircraft than from ships to correctly assess the situation and it is not always possible to intervene in the actions of aircraft when they are in the air. According to the pilot their craft over the island of Ibis were fired from a battleship. In response the aircraft dropped bombs on him that hit the stern of the ship. The number of victims on the "Deutschland" reached 80 men. When the squadron, fulfilling its main mission, was moving towards the transport, the radio of all countries broadcast sensational news about the "attack" of Spanish aircraft on a German battleship. The "Deutschland" went to English Gibraltar, passing there a request to provide an opportunity to repair the ship, as well as to prepare a certain number of coffins "to send the remains of the dead to their homeland." Feverish communications were going on between the German ships. Something evil was afoot. General Douglas who arrived from Murcia this evening was very interested in the situation and discussed with me precautionary measures. I recall that the tension was so great that the possibility of Germany declaring war on Republican Spain was not ruled out.

The next day was calm. Only on the air was the episode discussed in every way and various assumptions made. Cartagena was a heavily fortified base and was ruled out as a possible target for attack by German ships and one could only speculate about other ports and coastal cities.

On the night of June 1, when the "player" was already successfully met and under the guard of the entire squadron and moving to Cartagena the Republican fleet met with German ships. The darkened Republican squadron unexpectedly ran into a formation of German warships. It was the battleship "Sheer" and three destroyers. A fully illuminated German squadron gathered at a rendezvous point. Captured by surprise, German ships quickly raised their national flags and lit them with searchlights. But nobody was going to attack them. Both squadrons quickly dispersed.

The next morning when the republican fleet entered its base German ships appeared at Almeria and without any warning began a barbaric shelling of the city. As a result several dozen buildings were destroyed and many civilians died.

This ended the episode with the battleship "Deutschland".[43]

The tragic episode with "Magallanes"

After unloading, the "Magallanes", as usual took goods mainly lead and fruit, for delivery from Spain to the Soviet Union. It was also decided to send on it a group of Soviet pilots and tankers returning to their homeland. Some of them have not yet fully recovered from their wounds and needed additional treatment.

On the appointed day, in June 1937, the transport went to sea under the protection of a destroyer. Before dawn already 40-50

---

43 Author's Note. The episode with the battleship "Deutschland" reflected on the work of the Committee on Non-Interference in Spanish Affairs which was formed in September 1936 in London with the official purpose of monitoring the exact implementation of the non-intervention agreement by all its participants. In fact, this Committee, mainly due to the efforts of England and France pursuing a policy of "appeasing" Hitler and Mussolini, has become a screen for German-Italian intervention on the side of Franco. The USSR waged a stubborn struggle within the Committee and widely exposed the true nature of this institution to world public opinion. Thus, he rendered considerable service to the Spanish Republic. It was under pressure from the USSR and all people of goodwill in other countries that the Committee in March 1937 was forced to adopt a plan that, to a certain extent, constrained the interventionist activities of Germany and Italy.

Taking advantage of the Deutschland incident, Germany and Italy declared their refusal to participate in the established control and regained full freedom of action. However the governments of England and France with the support of the United States did not want to abandon the Committee and, sacrificing their pride, begged Germany and Italy to return to the Committee.

miles from the African coast the destroyer wished "Magallanes" a "happy journey" and decided to return to base. Subsequently, the territorial waters of Africa were to serve as the only protection for the transport. But then an unexpected misfortune happened. Dashing but ineptly maneuvering, the destroyer did not manage to slip past the bow of the huge transport and was rammed in the side. Deprived of the ability to move, with a big list, the destroyer was far from its base, endangering both itself and the transport. Having lost contact with the ships the fleet commander first sent a reconnaissance plane there. Arriving at the scene of the accident the plane could not figure out the situation and mistakenly reported that the transport as a rebel cruiser. Subsequently, already in a calm atmosphere our comrades who were on the "Magallanes" told how republican Katyushas, that is, the SB, flew up to the transport opened bomb hatches and seemed ready to drop their deadly load. "We even waved our suitcases on the bridge to show that we were here," they said, laughing. Fortunately the pilots, who were ordered to be careful (so as not to bomb a neutral ship), did not drop bombs on their destroyer and the "enemy cruiser", but wrote a note and dropped it directly on the bridge of the ill-fated transport. The note said: "If the warship is republican, let it sail around the "Magallanes", otherwise ..." But the whole trouble was that the destroyer could not move and therefore could not prove that he was our own. Fortunately, by this time it was possible to restore communication with the transport and at the very last moment return the planes to Cartagena. To save the transport four destroyers were dispatched from Cartagena, which having arrived at the site finally established the truth. They secured the ill-fated ships and brought them to the base. However as they say trouble does not come alone. "Magallanes" having again sailed to sea and safely reached the Dardanelles managed to collide with other republican vessels there, received serious damage and was under repair for a long time. It was hard to say what was the fault here - illiteracy, carelessness or malice. In any case, this was the last voyage of the "Magallanes" and at the same time the last attempt of the republican government to obtain military supplies from the Sevastopol - Cartagena route.

The balance of forces at sea by the end of 1937 had too changed in the direction unfavorable for the Spanish Republic and it became impossible to use communications through the Mediterranean Sea. It was necessary to resort to another way - from the Baltic to French ports of Le Havre and Cherbourg and from there by rail through France. Such a route was safe for ships delivering weapons from the USSR but it was very politically unreliable: transit through France depended entirely on the rapid change of government and situation in the Third Republic. Arms that came to Le Havre or Cherbourg were either let through or delayed and under any conditions a very greasy "lubrication" of the French political and administrative apparatus was required to deliver goods to Spain.

This dependence on Paris in the military supply of the Republic played a fatal role at the last stage of the war: at the very height of the battle for Catalonia (December 1938 - February 1939), a lot of weapons accumulated on the French-Spanish border, more precisely on its French side (aircraft, guns, tanks, torpedo boats, etc.) purchased by the Spanish Republic in the USSR and in other countries; if these weapons were in the hands of the Spanish army the whole course of the battle could be different but the French government despite all the demands and requests of the Republican government of Spain refused to give out its military equipment. This contributed to the defeat of the Republican army in Catalonia and consequently, the final victory of Franco and his fascist patrons.

Naval operations

Although the most important task of the Navy in 1936-1937 was the safeguarding of sea transportation from the USSR to Spain, yet the matter could not do without military operations against the rebel fleet. First, escorting transports coming from thc East sometimes naturally escalated into battle at sea. Secondly, republican sailors were eager to fight with the hated fascists, often because there was a lot of talk about the imaginary "inactivity" of the fleet in the rear where people knew little about the heroic efforts of the fleet to ensure military transport from the USSR. Therefore, the sailors used every opportunity to engage in battle with the rebel ships or to fire at the ports and bases captured by the enemy.

In April 1937, in the interval between the arrivals of the next "players", the commander decided to conduct a series of active operations against ships and enemy bases.

Chained to base after the northern campaign I asked for permission to take part in the planned operations. Consent was obtained, and I moved to the cruiser "Libertad".

In early April it was decided to join the cruiser and 6 destroyers in search of rebel ships which often went to intercept our transports. In the absence of rebel ships, the formation was to shell Malaga and the African port of Melilla.

According to s already established procedures, destroyers plowed out of the base at full speed in search of submarines as the cruisers slowly turned around in the harbor and went out to sea.

Hiding their intentions the ships headed north and only after dark turned south. Now in full speed they were moved on a broad front in search of the "Canarias", but to no avail. We limited ourselves to sending destroyers to shell Malaga and Melilla and return to the base by the evening of the next day.

On the cruiser "Libertad" they were unhappy. "Only anti-aircraft guns are firing, and the main caliber has not fired a single shell for more than half a year," complained petty officer F. Mira.

In mid-April it was decided to conduct a larger operation involving a battleship, two cruisers and a destroyer flotilla. This operation has taken place and deserves a brief description.

At that time, the situation at sea was as follows: the new rebel cruisers "Canarias" and "Baleares" went in operation and based in Cadiz. They occasionally appeared in the Mediterranean Sea, but were afraid of open clashes with the republican fleet. Despite the superiority of these cruisers in speed and power of artillery the Republican squadron did not avoid meeting them. Everyone was sure that the personnel on the rebel ships did not want to serve Franco and would not fight against the Republicans. There were many reasons for this. The rank and file of the fleet during the days of the rebellion on all ships without exception remained faithful to the legitimate government of the Republic and it took Franco a considerable amount of time to make a serious change in the composition of the ship's crews.

On this basis was conceived the plan of operations: the Republican fleet in full force attacks the coast in the area of Malaga - Motril and in the case of the rebel cruisers appearance to force them to accept battle. The operations of the republican fleet in the indicated area would confirm its supremacy at sea and calm public opinion in the Republic.

In the evening of April 15 a squadron consisting of the cruisers "Libertad" and "Mendes Nunez" and accompanied by a flotilla of destroyers left Cartagena. On the morning of April 16, battleship "Jaime I", which had left Almeria, joined her.

In the afternoon of the same day the ships approached the enemy shore and began shelling the port of Motril. A group of destroyers allocated for the shelling of Malaga was located to the south performing its task.

Approaching, in a line ahead formation, 40-50 cables (4-5 miles) to shore, cruiser the "Libertad" opened fire, and after it in the prescribed manner, fired the battleship "Jaime I" and the cruiser "Mendes Nunez".

I wanted to get to closer acquainted with the actions of the destroyers and so I moved to "Antequera", where the flotilla commander Vicente Ramirez and V. Drozd were located.

Following the head of the column we could perfectly observe the fire of all the ships. Besides weak resistance in the Motril region where a medium-caliber battery opened belated fire there was no opposition. Ships, as at the training ground, in turn opened fire on some object on the shore. The battleship and cruisers were limited to several salvos so as not to consume a lot of ammunition, and the destroyers fired a more massive fire. With the short distance to the coast and excellent visibility it was possible to observe the fall of each shell. Individual fires occurred after the shelling but no explosions of the alleged ammunition depots were observed.

About 15 hours on the horizon appeared first smoke and then the silhouettes of our destroyers. Breakers under the stem and astern indicated that they were in a hurry to join the squadron but why was not yet clear. Perhaps the commander of the fleet had already received a report from them, and we on the "Antequera" did not know anything yet. In answer to my question, don Vicente threw up his hands, swore at someone unknown - "Caramba" - and ordered something to the signalman on the bridge, apparently trying to find out the reason for the early return.

Soon everything became clear. Several rebel torpedo boats, keeping a certain distance, tried to attack the destroyers. The latter occasionally fired on boats. The boats departed but as soon as the destroyers turned to towards shore to shell Malaga the boats again attacked and the destroyers abandoning their intention to shell the base fired on the boats and it all started all over again. A kind of cat-and-mouse game. The destroyers, possessing great speed, were not afraid of boats and showing them their stern, went to sea. The boats were not afraid of the destroyers and when they opened fire on them, retreated to their base - Malaga. Approaching Malaga with the whole squadron we could observe this picture,

Then several rebel aircraft attacked the squadron. All this as well as the approaching evening made the commander expedite the return. At about 4 pm the squadron turned north. The destroyers, "playing" with the boats, took their place in the ranks.

It was difficult to expect great results from this squadron action but the mood on the ships was combative. There were lively conversations on the upper deck, sharing their impressions of the alleged losses from the enemy. Ramirez, all the while running from one side to the other and even rising to the top observation post, sat down on a round swivel chair took off his cap and wiped his forehead, saying: “Muy bueno, muy bueno.”

With the onset of darkness, the squadron continued northward reforming into line ahead and reducing the distance between the ships. The battleship “Jaime I” accompanied by two destroyers headed for Almeria. Cruisers and destroyers followed to Cartagena. Now in the head of the column was the cruiser “Libertad”, followed by the “Mendes Nunez”; the destroyer "Antequera" was the third, leading the flotilla of destroyers.

All personnel were required to be at their combat posts. But fatigue was telling, and the men were sleeping under their pea-coats in the recesses of the upper deck. The southern starry sky without a single cloud resembled the sky above Sevastopol. In April the Black Sea is warm only during the day and at night and especially in the predawn hours it is so cold that you have to put on an overcoat on the bridge; in the Mediterranean Sea you can sit all night in a light tunic. This blessed sea! No wonder I have often heard from Spanish merchants sailing across the ocean to America that when they enter the Mediterranean Sea, they say that "I have already arrived at port." This, of course, is not entirely true. Storms occur here but they are rare and endured much easier than in the northern latitudes.

As I sat on the bridge scanning the horizon and the ships with my binoculars, I thought about how fate had suddenly thrown me here. Longing for my own fleet instilled a melancholic mood. Remembering my cruiser I tried to imagine what he was doing now and what changes had occurred in the personnel during this time.

When the dawn began and the lunar path disappeared in the light of the coming day, the squadron had only a few hours to get to its base. However, these remaining hours turned out to be rather restless for fleet commander M. Buiza. The first unpleasant surprise was the message that the battleship "Jaime I", entering Almeria with the lights out ran aground and asked for help. And half an hour later an unknown plane flew along the starboard side as if portending the possibility of an attack from the air.

"The squadron must go to Almeria and guard the battleship," Vicente Ramirez expressed his opinion excitedly.

"But the more ship that are there the easier it is for the rebel planes to attack the squadron," answered his chief of staff, who was distinguished by an unusual calm for the Spaniards.

"What do you think, Nicholas?" - He turned to me, seeking support.

Not wanting to interfere in the dispute I replied that the commander knew better what to do. However, I personally consider it appropriate to go to the base and be in a state of readiness if the situation requires it.

Advising Vicente to ask the fleet commander about his intentions, I waited for an answer. "Libertad" was silent for a long time. Then a short message was received: "I'm going to the base." It seemed to me that the commander himself was in a quandary and hesitated for a long time until he made this decision.

The mood on the bridge deteriorated. The supposed joyful meeting in Cartagena was overshadowed by a major nuisance -

the battleship was aground. Suddenly other events complicated the situation.

First, the German cruiser “Leipzig” appeared on the horizon 40-50 miles from the coast which constantly stuck near Cartagena. He kept within sight on the horizon. But that was a bad sign.

Ramirez did not miss the opportunity to swear hard at the Germans but considering himself safe he invited me to drink coffee.

- Con cognac about con leche? (With cognac or milk?) - I asked him, jokingly.

- Hombre! He said and smiling he lead the way to the wardroom.

As soon as we managed to go downstairs an unusually loud and disturbing voice from the bridge: "Aviones!" Made us return upstairs. Everyone on the bridge scanned the sky with binoculars but no planes were visible.

“The destroyer “Lepanto” is being bombed by airplanes,” said the signalman and pointed in the direction where this ship was.

Indeed, pillars of water rose near the destroyer, and smoke from the funnel indicated it he was working up to maximum speed.

No enemy aircraft were visible ... Meanwhile, high columns of water rose in the immediate vicinity of the destroyer. What is the matter?.. Finally, we guess: these are not planes, they are ships shooting that are somewhere far at sea such that we don’t see them.

“Canarias!” Cried Ramirez.

His assumption was correct. The squadron turned towards the enemy. Increasing speed, the ships began to smoke heavily, their formation broken. The cruisers pulled ahead and the destroyers, clearly not knowing what to do, made dangerous turns, crossing each other's courses. M. Buiza did not consider it possible in the daytime, not possessing the necessary speed, to fight with the faster enemy cruisers and decided to send destroyers to the attack. The cruisers had to support them.

After 10-15 minutes the shelling of the "Lepanto" ceased. No enemy ships were visible. Having gone a little towards the supposed location of the rebel ships the commander turned back heading for base and canceled his order for the destroyers to attack. This decision was correct. The destroyers could not attack in the daytime a freely maneuvering enemy which had almost the same speed as them.

However, tactics have always been the weak point of the personnel of the ships of the republican fleet. Sailors rushing into battle grumbled at the indecision of the officers. Anarchists in such cases added fuel to the fire and hurled loud, irresponsible phrases.

Now I regretted the decision to go on this operation on destroyers and at the time of the complicated situation I did not know what the commander was thinking. The situation was really difficult. At the entrance to Almeria there was a battleship, still aground, that the destroyers, tugboats and even a submarine tried to pull away. He could easily be the victim of a skillful air raid or a bold raid by rebel cruisers. The destroyer "Sanchez" was enroute from Almeria to Cartagena which could freely fall under attack. And finally, the squadron did not have enough fuel to undertake a long operation at sea.

Here is Cartagena.

As usual, the cruisers were the first to pull into the harbor and the destroyers continued to carry out anti-submarine patrols at the entrance at high speeds.

"Where do you think Sanchez is now?" I asked Vicente carefully, as if hinting that he was supposed to worry about his destroyer.

"He should be arriving in Cartagena soon," he answered and shouted at the signal bridge to keep a good lookout there.

"Or maybe you should wait for him before entering the harbor?" The deputy recommended to his senior.

But no answer was required from Vicente. On the horizon, releasing black puffs of smoke and raising a breaker with its bow, a destroyer rushed at full speed to Cartagena. For several minutes we could watch shells bursting around the destroyer and bursts rising above the funnels. It was obvious that these were the explosions of the 8-inch shells of the cruiser "Canarias" or "Baleares" and possibly both of them together. The destroyer continued to move forward as in a boiling cauldron.

- Hard a starboard! - Ramirez commanded and the ships turned one by one following their flagship.

It was clear to all that a single hit on the "Sanchez" would almost certainly sink her or make her a stationary target for the enemy.

It was even less tactically sensible for the four destroyers to meet the cruisers now than it had been when they were at sea, but it was impossible not to go to the aid of a sister ship in distress. Everyone stared with bated breath at the destroyer moving through the explosions. The flotilla was moving forward, accelerating to full speed.

Hitherto invisible even through binoculars, the enemy cruiser was now visible on the horizon, first as funnel smoke and then gradually as a silhouette loomed on the horizon. The distance decreased: 190-180-170 cable. At the closest approach of the enemy's cruisers, the 15-inch Republican battery at Cape Cinezas opened fire and fired two or three volleys. It was the most intense moment. But the shell bursts around the "Sanchez stopped". The distance between "Sanchez" and the rebels began to increase rapidly. When the range-finders marked 195 cables only individual smoke caps indicated which direction the rebel ships were heading.

The commander led the formation to where the rebel cruiser was last seen, increasing speed to its fullest. High columns of smoke rose above all the ships, and the shore gradually receded behind the stern.

- Muy bueno, muy bueno! Shouted Don Vicente Ramirez, waving a radiogram. It turns out that a report was received that the stranded battleship had been refloated without damage and should be on its way to Almeria.

As quickly as the situation had become more complicated a few hours ago, it was now clearing up for the better. The “Jaime I” and 2 destroyers were finally safe in Almeria. “Sanchez” safely joined the squadron. Enemy cruisers retreated in an unknown direction. As it turned out later the “Canarias” and the “Baleares” using German intelligence decided to approach Cartagena and, in case of meeting with individual ships of the Republicans, attack them. Obviously to lure the enemy, one cruiser - “Canarias” - approached the squadron and the other kept a little further out ready to support him.

- Donde estan los aviones (where are the planes, where is our aircraft)? - repeated several times Ramirez.

I was also interested in this question. It seemed that it was impossible to come up with a more convenient opportunity to attack the ships approaching the base. But the planes, of which the fleet had only two units (5–6 medium bombers), bombed targets on the flank of the army and having just returned, could not still fly into the air. What an opportunity was missed! But, when you have only a few planes, it is difficult to plan their flights for all occasions.

By evening all the ships were in their places at the base, taking fuel and preparing for the next operation. But talk of the events of the last two days, the battleship stranding, the fortunate event with the “Sanchez”, and especially the appearance of enemy cruisers directly at the base long served as the main topic of lively conversations and discussions on ships and on shore.

I would like to note the heroism of the personnel. It manifested itself in the very first days of the rebellion when courage and decisiveness was required to prevent rebel commanders from raising people to rebel against a legitimate government. In the land garrisons there was no such unity and political maturity - there the soldiers, as a rule, resignedly obeyed fascist officers.

Amazing examples of courage were shown by crews of ships that did not have the opportunity to go to sea at El Ferrol. Sailors and non-commissioned officers with rifles and machine guns of the old model took to the streets of the city and repelled the attacks of well-armed ground units of General Maul for several days.

The air garrisons of Spain acted in the same spirit. Unfortunately, Spanish aviation was very weak at that time. A few dozen old French-made Potez aircraft and literally a few more outdated Dornier seaplanes - that's all that the Republic owned. Ignacio Cisneros, whom I mentioned in the early days of the uprising, got up on an old plane and bombed the Montaña barracks in Madrid which the rebels occupied.

In history, it often happened that in difficult times for their country, sailors sought to be where they were needed. This happened in Spain. When a decisive battle for the capital broke out on the outskirts of Madrid, a marine corps company was organized. She was thrown to the most important site in Carabanchel and Usera and fought alongside the famous Fifth Regiment. This company has repeatedly repulsed the attacks of Moroccan cavalry and Italian tanks. All Spain knew about the feat of sailor Antonio Coll. It was he who, at the most critical moment of the advance of the six Italian tanks, hit the first tank with a grenade, causing confusion in the others. The enemy's attack was stopped.

Sailor Coll died, "stitched" by a machine gun burst but following his example others rose and sacrificing themselves, stopped the enemy. This heroic example recalls numerous episodes on the fronts of the Great Patriotic War - in Odessa and Sevastopol, Libava and Tallinn, on the rocks near Murmansk and in parts of naval brigades on the outskirts of Moscow. Just as we had been imitated by commissar Filchenko, the famous hero of Sevastopol, so in Spain, special units formed to fight tanks were named after sailor Kohl.

From the mouth of eyewitnesses I had heard a lot of how boldly surface and submarine crews acted during the first squadron operations in the Gibraltar area in July - August 1936. The personnel not only did not fear meeting with the rebel ships, but also disregarded the danger during air raids when the anti-aircraft defenses of the ships was clearly weak. On the battleship Jaime I, a bomb smashed the upper deck, there was a threat of internal explosion, but the crew bravely coped with the fire and asked the Committee (there were no commanders yet) to continue the operation until the end. The actions of the destroyer "Diez" can serve as a model of devotion to the motherland. After bold operations at the beginning of the war in the Mediterranean, he was left in Cantabrico to protect the northern coast. Together with the destroyer "Ciscar" he conducted a series of bold operations. Later in 1938 "Diez" made repairs in French port, not succumbing to the numerous provocations of the rebels.

This is also a kind of heroism, and when the destroyer had to break through Gibraltar at the end of 1938, despite the failure of this attempt, both the personnel and the commander of the destroyer, merchant sailor Jose Antonio Castro, fulfilled their duty to the end.

Later, when in Moscow I was told about the sinking of the rebel flagship "Balears" by the destroyers "Sanchez" and "Antequera" in March 1938. Despite major shortcomings in the operational and tactical order, the personnel of the destroyers acted boldly and selflessly.

The commander of the submarine flotilla, R. Verdia, told how bravely the submarines acted during the days of the rebellion, how he himself attacked the rebel battleship "Espana" and how much the boats he could have made if they had a good material and trained officers. Later, I heard a lot about the actions of the

"C-6" submarine in the Bay of Biscay. When the damaged boat could no longer operate, it was scuttled, the crew reaching Saint-Nazaire in France, repaired another boat and returned to the Mediterranean Sea ready for further struggle.

Fighting for communication escalates

By the spring of 1937, the situation at sea had changed. With the arrival of two new rebel cruisers based on the island of Mallorca in the area of Cartagena the war at sea entered a new phase. It was not difficult to guess that the main attention of the enemy will now focus on sea communications and above those connecting Spain with the ports of the Black sea where the weapons came from. It became obvious that it was no longer possible to limit the allocation of several ships to meet the "players". It was dangerous to set a rendezvous point somewhere near Cape Tenes off the coast of Algiers and leave the transport route from Bizerte to Tenes undiscovered. Meanwhile, the transport Santo Tomé, which was also the largest, was already ready to leave and I was required to answer urgently to Moscow, at which point and which ships would meet it. The cargo on this ship was too valuable to risk. So far we have not lost a single ship with military equipment and even more so it would be unacceptable now that the formation of new units of the people's army (after Guadalajara) was entirely dependent on incoming cargo. The civil war had been going on for about a year. From occasional skirmishes between ships both sides moved on to more systematic combat operations. It was no longer possible to ignore the presence of rebel cruisers, as was done in September – October 1936 during the Northern campaign.

The advantages of the first months of the war when all the personnel except the officers were on the side of the Republicans were less pronounced at this time. Although the moral factor still played a huge role, success in battle depended on the power of the ships and the ability to use their forces. On the agenda was the question of strengthening the Republican squadron with submarines, torpedo boats and aircraft. The experience of the previous months showed that M. Buiza would boldly lead the cruiser into battle, on which he holds his flag commander, but hardly be able to manage a complex naval battle. And it wasn't just up to him. With low literacy and training, it is enough for the commander of one ship to do the wrong thing, make an erroneous maneuver, and it will confuse all other ships. The idea of the interaction of the surface fleet with submarines and aircraft suggested itself.

I went to the "Libertad" to talk with the commander, to find out how he assesses the new situation in the Mediterranean Sea and how, in his opinion, it is better to organize a meeting of the next "player".

I happened to be at a dinner where all were assembled. Even the destroyer flotilla commander V. Ramirez was here. Buiza invited me to the table. Having poured a large glass of red wine he raised a toast to all "Kompaneros Rusos", we drank. The Spanish comrades continued the conversation which I involuntarily interrupted by my arrival. Acquaintance for many months and sailing together on a cruiser and on destroyers brought us closer and we met as good friends without any ceremony. Having caught the thread of the conversation, I realized that the latest events at sea and the successful completion of the operation were discussed. The destroyer flotilla commander spoke in detail about individual episodes from the last operation in the Motril - Malaga area. He enthusiastically talked about how skillfully his destroyers acted when they were attacked by rebel torpedo boats. He convinced them not to be afraid of them and by correctly maneuvering one would always avoid torpedoes.

He became even more carried away when he described the breakthrough of his "Sanchez" under the fire of the rebel cruisers. We observed this ourselves but on the basis of what he heard from the "eyewitnesses" he now specified, "how several shells exploded so close to the destroyers that some sailors were knocked down." Everyone knows how, after various incidents, legends arise that overshadow actual events.

There was no harm in Ramirez's exaggeration and everyone listened to his story. Believe it or not. As usual the commander was mostly silent and only occasionally made comments. His chief of staff, L. Hunker, a spineless and frightened officer, was ready to carry out any order, but did not dare to have "his own judgment", even in this unofficial setting. The conversation's initiative was in the hands of the temperamental Andalusian don Vicente Ramirez, who, having finished the story about the destroyers, began to swear obscene words at the fascist, and then, pointing to the Balearic Islands, offered to raid the rebel cruisers based there. He was getting serious. He considered aviation to be the most suitable means of attack while the navy was assigned a modest role, and even that was mainly done by destroyers. "We need to ask the minister for airplanes," he said, looking in my direction, apparently wanting to show that this largely depends on me.

I did not want to get into the subject of my conversation with Buiza and I did not intervene in the debate waiting for it to end. Don Miguel looked at me and, obviously, understood that I had arrived for a certain purpose stood up giving the signal to the others. They began to disperse.

Already sitting in his cabin, I told him about the imminent arrival of transport and the importance of the cargo.

"We will go out with the whole fleet and meet him near Cape Tenes," said the commander. But this is what I was afraid of. I thought that we should abandon the old scheme of meeting transports near Cape Tenes, especially since the long way from the Tunis Strait to Algeria had to go without any cover for transport. That and the composition of the escorting ships needed to increase. It was no accident that I received a special warning from Moscow about the importance of the cargo and providing reliable security. I spoke in favor of a more Eastern meeting point. After a brief consultation, it was decided to leave earlier with the fleet, make a diversion before dark, and go East at night to meet the transport. I wrote down the latitude and longitude of the rendezvous with the "player" to report it to Moscow.

The remaining 5-6 days before the fleet departed was quite enough for maintenance, fuel loading and personnel rest. We sat for a long time and discussed various naval problems. I was most interested in the commander attitudes to the subject of using at sea available submarines, aircraft, and expected torpedo boats and to ensure maritime communications,. By the way, Buiza was delighted when found out that torpedo boats would arrive with the next transport. Understanding the role of aviation he asked me to increase the number of aircraft assigned to the fleet. Don Miguel was skeptical about the use of submarines in conjunction with surface ships. Yes, by this time they were not in Cartagena.

Very carefully so as not to hurt the fleet commander's pride, I asked what, in his opinion, would be best course of action if the republican squadron met both rebel cruisers. Buiza said that he is not afraid of such a meeting and would attack the enemy ships from different directions by cruisers and destroyers. Looking ahead, it should be noted that he so acted when such a meeting took place in September 1937, but, engrossed with the battle with "Canarias" and "Baleares", he did not fulfill the main task: to protect the transports.

We went out to the upper deck. Spring was felt in the air. The mountains, not having time to burn out, turned green and the city did not look as gloomy and dusty as in the summer. Ships stood near the pier and in the arsenal. The staff was preparing to go to sea. A new, more complicated stage of the struggle at sea began. The role of our sailor volunteers increased. They got comfortable with working on ships and without them I could not do anything. According to the established rule, I invited our advisers from the cruiser, destroyers and Ramishvili to my meeting. Much depended on their skill and experience, especially now that the entire fleet was preparing to go to sea.

While on the bridges of ships, our advisers unwittingly became participants in the fighting, their recommendations in an environment that requires the rapid adoption of opinions were often both advice and orders. I saw the flotilla commander and our adviser V. P. Drozd on the bridge during the operation to bombard the coast and Malaga. They calmly discussed the situation when the enemy was far away and when the "Canarias" appeared on the horizon, when every second took on significance, V. P. Drozd only pointed with his hand where he would go, and Vicente, if he agreed, immediately gave the necessary orders. On submarines the adviser became the commander and there he had to simply give orders. His position, it must be said, was very difficult. It required restraint. The fate of the boat often depended on it. So the operations of the Republican fleet became more complicated every month

They quietly discussed the situation, when the enemy was far away, but worth it appear on the horizon "Canarias", when every second becomes important, VP Drozd only shows his hand wherever he went, and Vicente, in the case of agreement, immediately gave the necessary orders. In submarines, the adviser became a commander, and there he just had to give orders. His position, it must be said, was very difficult. Exposure was required. The fate of the boat often depended on this.

So, the operations of the republican fleet became more complicated every month; when discussing the plan for the game to meet with our Soviet comrades we involuntarily talked about this a lot. Enough has been said about the actions of surface ships. I would like to tell you how the aircraft attached to the fleet, the republican submarines and torpedo boats fought. For this, I use not only the facts that I have witnessed, but also eyewitness accounts.

Spanish submariners

In the 30s, not wanting to lag behind fashion, Spain with the help of Italians acquired submarines. First appeared small type “B” boats and then newer type “C” ones. Partially built in Italian shipyards these boats were armed with torpedoes made in Fiume

(Italy). Their fighting ability depended on Italian equipment. This was especially felt during the war years.

After the overthrow of the monarchy the naval politics of the republican government was more oriented toward England. In the Spanish fleet the British Admiralty dominated the rebellion. English capital acquired the shares of shipyards in Cartagena, Ferrol and even managed to administer some of them, say, "Constructor Naval", transferred to London. Therefore, the newest type "D" boats are laid in Spain with the help of England. Don Patricio Casado, head of the Arsenal in Cartagena, was once examining the unfinished hulls of the large "D" boats, and told me that he could not work on them without the knowledge of his superiors in London and strangely enough in the first months of the war it was so. The English consul who was at that time in Cartagena freely visited the shipyard and gave "advice." True, this did not look so wild as the presence of the German consul there when Germany had openly helped Franco. However, we will not shy away. By the beginning of the rebellion there were 12 submarines in the ranks with the exception of one small one that was under repair. The condition of the submarines was much worse than the condition of surface ships, on which the nobility has long served and which had many good traditions, although they have been used more recently for ceremonial voyages than to prepare for hostilities. Submarines did not have clear operational tasks and at the same time were not suitable for parades in Santander during the king's stay there. The boat officers were drawn into the plot just like the surface ships officers. Most of the boat commanders went over to the side of the rebels and their place was taken by young officers, sometimes only by the will of fate remaining on the side of the republican government. Volunteers from the Soviet Union and volunteer sailors from France and Yugoslavia rendered great assistance to the republican submarine fleet.

When I first arrived in Cartagena in September 1936 I was introduced to the submarine flotilla commander R. Verdia. This man, as I wrote above, played a huge role in the fate of all the other boats. Short in stature, thick, with dark eyes, he stood out among other Spanish officers by the lack of excessive expansiveness. He spoke slowly and calmly, which often happens to people when they are in their place and know their job well, confident in themselves.

One evening the commander of the base gathered several ship commanders including Verdia. This time lunch did not last long. Everyone was in a hurry somewhere and Verdia remained to spend the night in the base commander's apartment and I got the opportunity to talk with him. Poor knowledge of Spanish did not allow me to understand much then. Gradually, from other sources, I restored the whole picture of the action of submarines, which I want to briefly talk about.

R. Verdia at the time of the rebellion commanded the submarine "C-5". All boats were based on Cartagena and Mahon (Menorca island). By the time the insurrection began they were in their bases and only one or two were engaged in combat training at sea. This greatly helped the ship crews deal with the rebel officers.

I knew that R. Verdia, commanding the "C-5", sailed in the North in the first weeks of the war - in Cantabrico. He attacked the rebel battleship "Espana" and was said to have even hit him with a torpedo but it did not explode. With this, I began a friendly conversation with Verdia. "Isn't it a shame when a boat risks itself attacking the enemy, gets a hit, and the torpedo does not explode?" I asked him. Any other Spaniard with his temperament would have surely burst out with regrets and curses at the torpedo and the old authorities. "Que lastima, que lastima" (what a pity, how sorry), - my interlocutor repeated several times, without even raising his voice. I sympathized with him. Indeed, it is a shame to seek and stay in position for the enemy a long time,, then, as a well-deserved reward, meet the largest rebel ship, the battleship "Espana", maneuver around it to launch an attack, calculate the lead angle for a torpedo salvo and without breaking away from the periscope finally fire at the right point to order a "salvo." Having fired torpedoes the boat usually goes deep and the commander nervously counts seconds in anticipation of the results. The friendly "cheers" by the crew of the boat meet the explosion of their torpedo, often accompanied by an even stronger explosion of the attacked ship. A blast wave indicates success. If the explosion was not heard, everything starts all over again. In this case, the opponents dispersed literally "out to sea." But the fate of both was unenviable. The battleship was soon blown up by a mine and sank in the waters of the Bay of Biscay. Submarine "C-5", when surfaced and for some reason unable to dive took on an unequal battle with two armed rebel trawlers and sank with all her crew. When the commander was asked how he was convinced that the torpedo hit but didn't explode, Verdia replied that he didn't go down to the depths and that he was thrown to the surface after the salve and watched the track of the torpedo. "The battleship was not dangerous to me," he continued (although this is not entirely true). In fact, it seems to me, given the nature of Verdia, he simply did not think at that moment about the danger in anticipation of the results. I didn't believe that you could see the trail of a torpedo going at a depth of 3-5 meters, but two weeks later, when I was on the "Libertad" bridge, at the entrance to the northern port of Gijon I could see how large fish "attacked" the ship and their footprint in the clear water of Cantabrico was perfectly visible at a depth of several meters. "Torpedo!"- inexperienced sailors shouted alarmingly several times then from the signal bridge until they got used to the situation. These were, apparently, dolphins with which our Black Sea sailors are well acquainted but on the Black Sea the water is darker and dolphins can be seen only when they jump out of the water in packs pursuing ships. "What clear water," I thought, standing on the bridge, and remembered the conversation with R. Verdia.

In the first days of my stay in Spain I wanted to know more about the state of submarines and the training of personnel. As a patriot of his profession Verdia clearly overestimated the fighting efficiency of the boats and the loyalty of the remaining officers. According to him, "the boats will deal with the fascist ships," and soon he will prove this as commander of the flotilla. I wanted to believe him but in fact this did not work. Verdia himself soon died and the boats lost an energetic leader. It happened like this: he went to Malaga, where the submarine "C-3" was lost under mysterious circumstances. Verdia wanted to find out the reasons for this misfortune. The boat left Malaga and exploded before submerging. They said that this was the work of an Italian submarine but it was equally likely to suggest an internal explosion. A boat is likely to turn on its crew and especially when transitioning from peacetime to combat conditions. After the explosion “C-3” quickly went to the bottom and only two people - the commander and navigator, standing on the bridge, were thrown overboard by the shock wave. They tried to swim to the shore. Having reached the coast, the commander fell down in a faint and died without regaining consciousness and only the navigator survived. In the same place, in Malaga, R. Verdia died during a bombing. Verdia spoke in great detail about the actions of boats in Gibraltar during the first weeks of the civil war. Having dealt with the rebel officers, the boat crews took part in the struggle for the Republic. All operating ships were on the side of the government, until the rebels were able to put into operation the cruiser “Cervera” and the battleship “Espana”. The main task of the fleet at that moment was to prevent the transfer of rebel forces across the strait. It’s easy now to retroactively talk about mistakes and missed opportunities At the same time, left without commanders, Spanish submariners, having barely brought their ships into combat-ready condition, did what they considered most important. Their combat effectiveness was, let's face it, small and their casualties were high. Events developed as follows: very soon the command was convinced that there was not insufficient activity for submarines in the strait and decided to transfer them to the North. What happened there where several provinces remained loyal to the government?

As soon as the rebels put into operation the cruiser "Cervera" and the battleship "Espana" they used them to block the northern provinces from the sea. After the fall of Irun and San Sebastian in early September 1936 only maritime communications could supply arms and food to the surrounding Asturias and Basque country.

At the end of July, rebel ships fired for the first time at the Republican coast and the North needed help. Sending boats there was the right action of the local command. But very soon the poor condition of the boats and the poor crew training was discovered. In the very first months of the war 3 type "B" submarines sank. Under mysterious circumstances, submarine "B-5" disappeared and it is possible that it fell into the hands of the rebels. I was not able to find out what really happened with this ship during my stay in Spain.[44] In August, being on guard at the Straits of Gibraltar, the submarine "B-4" sank with all its personnel. On September 19, the destroyer "Velasco" sank submarine "B-6" near Santander. This, in fact, ended the participation in the war of boats of this type. The rest were under repair and were declared unfit for work. Such is the fate of these old and small type "B" boats.

Type "C" submarines also did not have a significant effect on the course of the war, although there were opportunities for this. Boats operated mainly in the North. At first, two "C-5"and "C-2" boats were based there. As already mentioned, the "C-5" under the command of R. Verdia attacked the battleship, after which the boat laid up for repairs and its commander died. Soon after, a reinforcement of two boats was sent to the North: "C-4" and "C-6". At the time the squadron arrived in Bilbao there was only boat "C-2" in the North which was preparing for operations. As a rule the submarines acted selflessly but almost all of their heroic efforts were in vain.

Wishing to help the Spanish comrades I asked my superiors to send some experienced submariners to Spain. Around February 1937, I. A. Burmistrov arrived in Cartagena. He was charged with finding out the reason for the failure of the boats and, if necessary, to take command of one of them. There were two submarines in Cartagena at the time. Here our volunteer submariner rolled up their sleeves and began to work. At first he was indignant at the condition of the boats, and said that "with material in such a state of poor maintenance it is possible to sink

---

44 The actual cause is not known as of this printing. The standard work on subject believes the ship was scuttled by her pro-Franco commander in an act of suicide. See Gonzalo Rodríguez Martín-Granizo and José Ignacio González-Aller Hierro. Submarinos Republicanos en la Guerra Civil Española. Madrid 2003

in the Harbor," but this was not news to us, and he had to do everything possible under the current state of Affairs. I suggested that he choose any boat and prepare for combat operations. Especially since the cruiser "Canarias" then often sailed from the Straits of Gibraltar to the Mediterranean Sea and was sometimes even based at Palma on the island of Mallorca. I do not remember how long it took to prepare the boat, but on the appointed day and hour boat "C-1" under the command of Burmistrov went in the direction of the Balearic Islands. She was to attack the enemy at the entrance to Palma. A few days later, the boat returned to the base and the commander said the following: the boat successfully chose a place to observe the enemy cruiser but the next day found reconnaissance aircraft constantly "hanging" over her and then there were bombers which began to be a disturbance, dropping their deadly cargo. The commander tried to maneuver and having gone deep, to change position, but the same thing repeated. After inspection Burmistrov noticed that the boat stretches an oil trail on which the aircraft found it. By this time a serious hunt had already been organized for the boat. There was nothing to do but go back. "I tried to go deep," the commander explained – "but it didn't help." And it is clear that if the boat leaves an oil trail, it will necessarily float to the surface and such a measure as diving to a greater depth will not help. Returning, the boat was laid d up for repairs, simultaneously engaged in combat training.

After some time, Burmistrov received a more difficult and responsible task: to pass through the Straits of Gibraltar, find and attack the enemy in northern waters and at his base El Ferrol. The boat passed through the Straits of Gibraltar twice submerged but could not boast of success. The material aspects functioned poorly. One by one equipment failed and the boat had to return to Cartagena. It should be noted that our submarine volunteers found themselves in more difficult conditions than those who fought on surface ships. In submarines they acted as commanders not advisers, not knowing how foreign technology should function and were often forced to command through an interpreter until they mastered a dozen or two of the most necessary commands.

After the squadron returned to the Mediterranean Sea in October 1936 the destroyers "Diez" and "Ciscar" and three submarines remained in the North. It was these forces that Valentino Fuentes commanded in 1937. His predecessor, Navarro, betrayed the Republic and fled, taking, by the way, with him a certain amount of gold. He fled, however, in a hurry and left a suitcase in his cabin with things far from personal use: rings, watches, bracelets, etc. When the new commander was asked what to do with this suitcase, he replied: "A suitcase with things is the personal property Navarro and they should be transported to France, "that is, where the traitor escaped. This is the whole of Don Valentino (and not only him); who was both a "gentleman" who easily coexisted in relation to his enemy and a republican who sincerely hated fascism.

The rebels were rushing towards Gijon; in October 1937 this heroic city fell. Destroyers and boats were first based on Bilbao, and then moved to Santander. "Ciscar" was soon sunk and "Diez" went to France and was interned in Gibraltar at the end of 1938 while attempting to break towards Cartagena. Destroyers and boats, considering their main task was to protect the coast, sometimes broke through the blockade, delivering fuel and food to northern ports. But back to the submarines.

Of the three boats operating in the North, one had a particularly large share of trials. It was a boat "C-6". When I first arrived in Cartagena in September 1936, she was standing in the Arsenal undergoing repairs after duty in Gibraltar. We went up the narrow gangplank and down the bow hatch. A team was working in the torpedo room. Two torpedoes were being put in order, almost completely extended from the torpedo tubes. The boat was not clean, but what was much worse, the crew did not believe in their torpedoes. Everyone knew of cases where the torpedoes sank after firing, went wrong, evaded the set course, or did not explode after hitting the target. In addition, there were few torpedoes, and two tubes in the submarine were empty, not to mention lack of spares. These torpedoes were purchased in 1928 from the Italians and since then they were carried as dead weight without proper care. No torpedo adjusting or bleeding was done in the workshops. The team tried to put them in order after each case of torpedo failure when firing. I was well aware that Fiume torpedoes were used on surface ships of the black sea fleet, but I was not sure whether we had torpedoes for submarine boats. Just in case I decided to ask my superiors to send torpedoes but we really did not have them. A few months later the Republicans were forced to buy torpedoes from the Greek government, paying exorbitantly without a guarantee that they would work flawlessly.

We went into the diesel compartment and climbed from the control central to the bridge tower. Crates of food and ammunition for a 3-inch anti-aircraft gun were piled on the upper deck. The boatswain was in charge of the eldest boat.

A few months later M. Buiza came to me one day and told me two boats were sent North including "C-6". I learned about her further actions from various sources later. In Bilbao the boat was damaged and lay quayside for several months. In July 1937 a new commander arrived - Matisse. It was a Soviet volunteer - Hero of the Soviet Union N. P. Egipko. I missed him. He arrived in Spain when I was already returning from there. He cannot be denied courage, and if to his account there is only one rebel ship sunk, this does not diminish his merits in those extremely difficult conditions. This is what I managed to hear about the actions of this boat. She several times unsuccessfully attacked the rebel cruiser "Cervera". After the fall of Bilbao in August 1937, all of the ships concentrated in Santander came especially hard days. The roads to Gijon were blocked and this made it difficult to evacuate the population. A lot of people gathered in the port. Everyone wanted to get on the ships. According to Egipko he had the difficult task of staying in the port until the last moment to load some valuables to be taken out of Bilbao, only after that he could go to sea. And then Don Valentino - the commander of the Cantabrico fleet - arrived aboard and demanded to leave port as soon as possible. The enemy was shelling the city. The crowd in panic anxiety rushed towards the ship. The boat, carrying out the instructions of the Basque government, nevertheless took on board a valuable cargo. It was no easier in Gijon where she arrived. The last port of Asturias was subjected to frequent and massive raids. Warehouses and fuel storages burned. Bombs exploded all around. In such circumstances, "C-6" repeatedly went out to combat operations and not without success. A gunboat was sunk. In critical days of October 18–19, 1937 the question arose: what to do next? At the commander's conference there were voices in favor of going to sea to save the ships. "C-6"'s commander insisted on fighting until the last opportunity. But he was out of luck. A large-caliber bomb exploded between the quay and the boat causing her permanent damage. I had to ask permission to scuttle the boat, so that it does not fall into enemy hands. But it turned out that no one has the right to scuttle a ship without the knowledge of the minister, and a confused don Valentino was afraid to take responsibility although time was running out. A committee was needed and Commander Matisse and Commissar Paolo gathered it to destroy the ship. The boat was sunk but what to do next? All the ships had already left and the situation of the city is hopeless. The population evacuated by an English transport which having gone to sea fell into the clutches of the cruiser "Cervera". An English cruiser came nearby and helped the refugees. He made it possible for the refugee transports to leave for France.

Soon Matisse was summoned to Barcelona to the fleet commander where he received orders to form a crew for the boat "C-2", which stands in Saint-Nazaire. Routine repair work in a peaceful French port turned into a real battle for the ship. The authorities formally help but in fact interfered at every step. Fascist thugs scurry around and agitate among the crew, offering a lot of money for betrayal. Among the personnel there were anarchists who slowed down the work. One sabotage after another unnerves and demoralizes the crew. In the end, the boat is ready for testing but it is not given the opportunity to go to sea. "Having gone to sea, you lose the right to return," said the port authorities. It was necessary to stealthy test the boat at night in the harbor basin and without properly checking mechanisms go on a dangerous voyage to break through Gibraltar to where the Republicans are still fighting. Almost in the same position was the boat "C-4" in Bordeaux. Both boats decided to simultaneously go to sea and proceed to their destination. This happened at the end of July 1938. With great difficulty, under the threat of discovery and liable to be sunk the boats nevertheless broke through the Straits of Gibraltar and came to Cartagena.

The longer it went on the more effort was needed by the commander to maintain the fighting morale on the boat. The officers obviously sabotage further operations and the crew is tired and demoralized by the retreat of the republican troops. Remembering the actions of the Spanish submariners, one considers over and over: "How much courage and determination there was and how little organization and success."

Torpedo Launches

There were no torpedo boats - a torpedo launch - in the Spanish fleet. Spain stretching clothes on its legs[45], was fond of light high-speed ships, relying on them even in case of a major war. "Why Spain did not build torpedo boats," I once asked base commander A. Ruiz. "We have long paid attention to the seaworthiness of ships and boats cannot be sent far into the open sea," he replied and there was some truth to that. Spanish ships were quite strong and seaworthy. They made great sailers and perfectly withstood fierce storms. However it seems to me the main reason was different. We must not forget about the influence of England on the Spain's maritime policy. There was no point in developing a weapon dangerous to large ships for a friend which can easily turn into an enemy and besides there was not much profit in it. In any case when I became acquainted with the Spanish fleet on the spot I did not find a single torpedo boat. It seemed strange to me, a Black Sea man: we were so used to seeing torpedo boats at all exercises that their absence among the Republicans was evident. The question of sending boats to Cartagena really came up a bit later. The reason for this was the relocation of the rebel cruisers to the Mediterranean Sea and the torpedoing of the cruiser "Cervantes" in the roadstead of Cartagena by an unknown

---

[45] Russian proverb meaning "according to their capabilities"

boat. Entries and exits to the port had to be guarded more and it was not profitable to use large destroyers for this.

In December - January, I asked my superiors and began clarifying negotiations with Moscow. Meanwhile in Cartagena, the question arose of protecting the water area and many small vessels were adapted for this purpose. An attempt was even made to install 45-mm torpedoes on one of these boats. The non-commissioned officer familiar with this kind of weapon turned out to be an enthusiast in this matter.

His proposal was treated with disbelief. In fact what could he build without engines and torpedoes? At the same time, I had the opportunity to meet one "inventor" who lived in Barcelona and called himself an anti-fascist and allegedly was planning to build one torpedo boat with his funds. He allegedly sailed on boats back in the years of the First World War and now "wanted to help the Republicans". I don't know how it ended up with the construction of the boat in Barcelona but the excessive "curiosity" of the "inventor" for the arriving Soviet boats forced me to be careful.

In the spring Moscow reported that the boats were "getting ready to be dispatched". In May 1937 I noticed from afar torpedo boats covered with tarpaulins were on the upper deck of the liner "Santo Tomas" entering harbor. These were boats of the G-5 brand with Isotta-Fraschini engines. Small and very fast, they are located on the deck of the transport. Their crews led by V. A. Larionov arrived with them. Of the commanders of the boats, V.P. Likholetov and A.P. Batrakov were also remembered. 4 boats arrived on this transport and a few more were expected later. So there was a small formation of torpedo boats.

The fleet commander was very pleased with this addition and agreed to personally deal with both the basing of the boats and the training of the missing personnel from the Spanish sailors.

Buiza proposed using Porman as a harbor for the boats, which reminded me of the small Balaclava Bay on the Black Sea with a narrow but deep entrance. Unsuitable for large modern ships, this small bay, sandwiched by high cliffs, on the banks of which lived workers of neighboring lead mines and fishermen, turned out to be very convenient for torpedo boats. Two small piers could fully satisfy all their needs. There was also housing. Better could not be wished for. Having decided to organize a base of torpedo boats here the commander invited a representative of the local government. An old man appeared with red watery eyes and timidly introduced himself to the fleet commander. While they were talking I decided to take a walk along the sea, on the shore of which a small village was located.

Quiet and abandoned (the mines were hardly working at this time), the village was completely without men, children surrounded the car and looked with curiosity at the sailors who came to them. They hardly visited Cartagena although they lived nearby. It was impossible not to notice that most of the residents are sick with trachoma like the old man who spoke with the fleet commander. I became interested in the causes of such a massive severe illness. “This is a common occurrence in lead ore mining villages,” driver Ricardo told me. “Now, if you could see at what is happening near Almeria ...” The republican government of Azana-Giral had not yet found a solution of these social problems. Then I often had to call in to Porman to my comrades and watch as dozens of people sat on benches near their houses, sick with this serious illness and seemed to be resigned to their fate. “The war will end, and the people will require a number of serious reforms,” I thought then believing that the Republicans would win.

So fare I had to think about the war. We agreed with M. Buiza that he will allocate the necessary number of specialists and seamen to serve Porman base. For a long time the commander of the Portman base, which served the boats, was A. P. Korobitsyn. As I have said when he came to Spain he worked as my interpreter and assistant but he always wanted to "fight" on his own and I had to grant his request.

Two weeks later, the "run in" of the boats took place. They were met with enthusiasm when they entered the bay at full speed where large ships stood. Huge Spanish flags fluttered on boats, and sailors on ships waved berets and as usual shouted "Viva la Republique!"

These boats could not operate far into the sea and the rebel cruisers did not appear at that time off the republican shores. Loaded with depth charges, the boats carried anti-submarine patrols at Cartagena, replacing destroyers. The weak side of the boats was their lack of seaworthiness and poor anti-aircraft weapons.

As they mastered the boats they were assigned other tasks. They escorted transports near the coast and went in search of the enemy not moving far from their base. From the stories of my comrades who returned from Spain after me, I learned that when the detachment was reinforced with new boats and coastal communications became more important the boats were used from mobile bases in other ports and small bays. For this purpose "mobile bases" were organized: several trucks, a gasoline tanker in this case accompanied boats in Alicante, Denia, Valencia, Taragona and other places of their temporary basing.

According to Katernikov who have gained experience in the G-5 in Spain these small, high-speed boats had a number of shortcomings in open sea theaters. "These are night-time weapons near their bases," the Katerniki later told me. It was impossible to disagree with him. The infantry-mounted Degtyarev machine guns delivered to them were already unsuitable for fighting modern aircraft. They did not work at a high elevation angle. Therefore when the boats found themselves in daytime without air cover they could not resist even such aircraft as the Dornier. A battle between boats and aircraft in October 1937 confirmed this. Three republican boats patrolled near Alicante at the place where a transport washed ashore. A Dornier arriving from Mallorca made the boats retreat, damaging one of them.

Wave surge and distance were also the reasons for the limited use of boats. So in March 1938 when the Republican fleet attacked and sunk the cruiser "Baleares" the boats could not go to sea due to rough weather and a wave surge of more than three points.[46] Currently completely different ships of this class are being built with excellent seaworthiness. Their sizes are now of considerable importance because it is impossible to count only on a calm sea.

Soviet volunteer pilots

I would like to tell you especially about consultations and joint work with wonderful Soviet pilots.

The Spanish Civil War took place at a time when aviation as one of the most modern types of armed forces was of particular importance in any large country. The lack of experience in using aircraft and especially at sea in the First World War did not prevent military specialists from drawing far-reaching conclusions about the role of aviation in the future. For war at sea, special large ships – aircraft carriers – were built, seaplanes were created, and the use of coastal land aviation in naval operations together with the fleet was studied. A number of maneuvers in the USA and Europe confirmed that any, even the largest, ship could be sunk with air bombs and this intensified the race for the construction of aircraft carriers and various types of aircraft. This even gave rise to one-sided theories in favor of the development of aviation, such as the American theory of Mitchell-Douai. However, the practical use of aviation both on land and at sea on a significant scale could be observed only during the years of the national liberation struggle of the Spanish people. The Pyrenees were a kind of testing ground for testing aviation equipment and the standards for operational use of aviation developed by this time.

---

[46] 3 points corresponds to a slight swell.

Spain did not have its own modern aviation. The obsolete "Potez" was in service with the land aviation and the Italian "Dornier" was the backbone of hydro aviation for servicing the fleet. Several marine airfields - San Javier, Mahon, Barcelona, Cadiz and others - served as bases for seaplanes. Organically, the Spanish hydro aviation was not connected with the fleet.

When the war took on a protracted nature, the Spanish Republican government felt an urgent need for a strong air force without which operations on land or at sea could no longer be imagined. Immediately after the start of the rebellion, Germany and Italy provided assistance to Franco precisely by aviation. So already in August 1936 formations of the Italian military aviation helped to transfer the foreign legion from Morocco to the peninsula. Then came German Junkers, Heinkels and Italian Caproni or Fiat on all land fronts. No major operation was conceived without aviation. Air dominance became a prerequisite for successful land and sea operations. This explains the urgent request of the Spanish government to help them as soon as possible with airplanes and especially fighters. The government was actually convinced that without fighter cover (and even with weak anti-aircraft artillery), the republican fleet could not be in the Cantabrico and Madrid (like other cities) felt helpless when raids became more frequent. Aviation was not used enough due to lack of aircraft and in cooperation with the fleet and the importance of such interaction was also undeniable. We can say that it was during the Spanish Civil War that for the first time aviation both on land and at sea took its proper place in military operations.

Aviation played a decisive role in the early days of the rebellion. It ensured the rebels a successful offensive where they would never have achieved this only by the ground forces combat, say, during the capture of Irun and San Sebastian. Finally, with the help of airplanes both sides conducted and supported large-scale operations. Therefore the actions of aviation in Spain were carefully studied in all countries and especially when political events showed the inevitability of a great war. We, the sailors, also learned a great deal from the modest experiences of interaction between republican aviation and the fleet. The defense of Cartagena - the main base of the fleet - for the first time forced the fleet command to turn to fighter aircraft as the most reliable means of repelling air pirates. As early as 1936, when rebel air raids on ships at sea were random and rarely reached their goal although the bombing took place at low altitude and only during the day, Cartagena experienced the power of air raids. We very soon became convinced that anti-aircraft artillery alone could not effectively repel the enemy, and the absence of fighters jeopardized the presence of warships in the port and operations to unload transports. Rebel aviation in September - October 1936 as already mentioned made it impossible for the republican squadron to stay in Cantabrico as a result of which it was decided to return the ships to the Mediterranean Sea, despite the fact that for operational reasons they were needed in the North to provide communications. A series of subsequent naval operations showed how important it is for the fleet to have its own aviation. Fighters were necessary to cover the actions of their ships and bombers supplemented the power of naval formations and created a threat to enemy ships.

When the republican battleship "Jaime I "was stranded at Almeria in April 1937 without fighter cover his position was critical. When the high-speed "Canarias" and "Baleares" appeared at the main base, threatening republican ships, the most real threat against them was bomber aircraft. In the battle for the safety of the transport convoys in September 1937 the Republican squadron could do nothing against the more powerful new rebel cruisers the role of aviation interacting with the fleet could not be overestimated. Dozens of such examples even from that peculiar small naval war that unfolded off the coast of the Iberian Peninsula showed that in a future war there can no longer be major operations at sea without aviation. We Soviet sailors too made many conclusions when we returned from Spain. On the one hand all this seemed to be not new because in our country in the course of peaceful combat training, great attention was paid to aviation for a long time and on the other hand, in Spain a lot of things came to light that could not be foreseen.

The use of aviation in Spain both on land and at sea fronts increased gradually with the receipt of aircraft and the scale of the civil war.

When in early October 1936 Cartagena underwent the first raids, both sides acted ineptly. The rebel planes suddenly appeared over the base hurriedly dumped their cargo and flew away. Bombs fell far from the target. There were no repeat raids that night. These were poorly resisted by the anti-aircraft batteries that fired inaccurately due to not having the means of detecting aircraft and having simple sights instead of good sights. Impacts of bombs on ships and shells on aircraft were purely random in nature. Minor objects suffered, most often not having any combat significance. I remember how one such raid found us in the premises of the base commander and we learned about it after the bomb exploded. No shelters were built then. We ran into the park when the plane flew over us at an altitude of 400-500 meters and we pressed ourselves against the wall of a building. For most this was the first raid. Only one comrade, who arrived from Madrid, lay face down on the ground and then assured us that in such cases this is the most reasonable: "I only survive danger when I see that the plane is flying over me and I'm lying still the rest of the time," he said. Nobody seems to follow his example. Everyone preferred to take cover at least in an ordinary basement. Well it seems to me this original itself soon abandoned his method of tanking shelter when more planes began to appear and they flew for hours over the base and at such an altitude that only the noise of the engines was heard in the dark southern nights, and the erratic explosions of shells and numerous multi-colored tracers from machine gun bursts covered the whole sky.

In November the activity of the port and fleet increased, raids became massive and cargo transports required serious security. Therefore the first fighters assembled in Los Alcazares took off into the air to cover the airfield and the base from rebel aviation. From this origin in fact the fleet began working together with aviation, with our volunteer pilots, which we cannot but help talk about.

Events developed in this way: fighters attached to the fleet in November and the installation of several new anti-aircraft batteries changed the air defense of the base for the better. The rebel planes were now cautious and flew at high altitudes. The conclusion made by us in this period of operations is that the most real and active means of defense is aviation. It became apparent that fighters designed to protect ships and bases should be subordinate to the fleet and act in concert with anti-aircraft artillery. Then we felt the need to have good means of detection in order for fighters to take off in a timely manner and meet the enemy on the way to the city. Only under this condition could our volunteer pilots reliably guard Cartagena. We owe it to them first of all that we did not lose a single gun when unloading transports.

As maritime operations developed the need to have at least a small number of bombers also increased. At first General I. Cisneros and General Douglas could not allocate a single one from the first batch of medium-sized bombers arriving in Cartagena then two or three months later, by March 1937, the republican fleet had five to six bombers.

For many months the republican fleet has been in contact with this small squadron under the command of the pilot I. I. Proskurov and navigator G. M. Prokofiev.

Clearly remembered are some episodes of joint work between sailors and pilots relating to the spring 1937. In these months, one after another in short intervals Republican transports came to Cartagena from the Soviet Union. The Spanish government asked us to speed up their unloading at the port and regardless of danger we did the work day and night. In addition, there was another reason for close cooperation with the aviation and flight personnel of Soviet volunteers. Tank units were formed and trained at Arcena, near Murcia and artillery and other weapons for the Republican army were sent to Albacete or directly to Madrid, the planes were assembled in the immediate vicinity of Cartagena. In addition sailors did not have close interaction with tanks or artillery and therefore almost never met with their personnel once they left port. It is true tankers or gunners came to

Cartagena but this was quite rare. Sailors had a different relationship with pilots. Sometimes a transport would arrive and somehow imperceptibly, without much difficulty, tanks or guns would be unloaded. Yes, this is understandable. They are placed on a truck or platform and are quickly dispatched from the port. But there is much more fuss with the huge crates holding aircraft fuselages. To help unload and transport this oversized cargo from the port to the airfields, there came to Cartagena not only pilots, but also engineers led by G. I. Alekseev, who was in charge of the assembly of aircraft and equipment. Sometimes the pilots were in Cartagena for two or three days and some of them felt at home here. Pilots came to Cartagena and just relaxed when given a free day.

Having met with I. I. Proskurov at the headquarters of the Cartagena base we decided to clarify what should be the assistance of aviation to the fleet.

I explained that the Republican fleet has three main tasks: ensuring communications, protecting the coast (in case of an attempt to land troops or shelling cities), and finally, finding and destroying enemy cruisers. "In all cases, your main adversary should be considered the large rebel ships," I finished and asked if the pilots could bomb moving objects at sea. "Well, I never had to deal with ships, but they taught us to bomb on moving targets, such as tanks," Ivan Iosifovich answered, not seeing yet the big difference in the actions of aviation over land and at sea.

In fact, everything turned out to be much more complicated. Well-trained and unconditionally brave pilots Proskurov, Ostryakov and others did an excellent job when it came to ports, railway junctions and other objects on land. But difficulties appeared as soon as the topic touched the sea targets, and the first was to recognize the ships at sea from a great height. This was compounded by the presence of neutral ships. For example, once we received information that two or three rebel cruisers were concentrated in Malaga then already occupied by the rebels. We decided to send airplanes there. Usually, such long-distance flights were made with the expectation of attacking either at dawn or just before darkness. Airplanes preferred navigating the sea along the coast. The question arose what to do if the ships were discovered not at the port, but at sea. There were ships of neutral countries near the ports being monitored by the Non-Intervention Committee and it was both easy to make a mistake and their inviolability had to be guaranteed. We agreed with Proskurov that he would confine himself to striking targets at the port until he learned to confidently recognize ships. The experience of training in peacetime on the Black Sea and the use of aircraft during the Great Patriotic War confirmed how difficult it is to recognize a warship from the air, when you cannot approach it without the risk of being shot down. Let us recall how at the end of June 1937 our volunteer pilots nearly bombed their vehicles and took the republican destroyer for the enemy cruiser. We observed the same picture during the attacks of the rebel aircraft on republican ships.

In March and April 1937, rebel cruisers began to appear frequently in the area of Gibraltar and the Balearic Islands, threatening sea lanes. Several attempts were made to bomb them at sea. Two such attempts ended unsuccessfully. "It seemed that they were hit, and the cruiser continued to move without visible damage," N. A. Ostryakov once complained to me. Now we know that successfully attacking a fast-moving cruiser with two or three planes from horizontal flight is almost hopeless, but we simplified this task earlier. It is a different matter to attack ships in the port. Recognizing them there is also not easy, but getting there is still easier. This was confirmed in the episode described earlier with the German battleship "Deutschland" on May 30, 1937. Then the pilot Khovansky, who replaced Proskurov, reported to me that he could not distinguish the German battleship from the rebel cruiser, and since he [the German] opened fire while in the rebellious port of Ibiza, he attacked him. And he was right! And the hit of one or two bombs in the stern of the ship confirms the ability of pilots to accurately bomb if they are dealing with a fixed target.

All these facts confirm the conclusion that aviation, called to act together with the fleet, must undergo special training. With insufficient pilots training, airplanes are equally dangerous for both the enemy and their own ships. I recall the case when the commander called planes from Cartagena during a battle with the rebel cruisers near Algeria in September 1937. Several times the planes appeared over the formations leading the battle and each time could not distinguish their ships from the rebel ones.

Later, after returning from Spain, I had to engage in the interaction of naval ship formations with aviation in the Pacific Ocean, I already knew what thorough preparation is required for naval pilots, and the Patriotic War fully confirmed this. At present, new technology has radically changed the forms and methods of battle, and yet war at sea is a very complex matter. Space, speed, and range of ships and modern weapons leave a certain imprint on naval warfare. But back to the Spanish events

I remembered the two pilots especially well, perhaps because on our return from Spain we continued to meet for several more years in the Soviet Union – these were Ivan Iosifovich Proskurov and Nikolai Alekseevich Ostryakov.

After graduating from a flying school with honors, I.I. Proskurov became an aircraft commander and then a member in one of the air units in the Ukraine when he had the opportunity to go to Spain. He was a courageous man. When he was offered to fly to bomb Granada or Malaga, he was more interested in whether there was enough fuel, at what time it was better to appear above the target, and least of all he gave thought about how dangerous it was. Together with navigator G. Prokofiev, he led his detachment to their maximum radius of action both on land and at sea. Once he told me that he became a pilot not "by desting since his birthday," as is sometimes said, but almost by accident. "I was even afraid of flying," he said, "but I was called to the Komsomol district committee and there they asked me to go to flight school." It seems to me that this is characteristic not only for him, but also for many famous pilots who ended up in a flight school as a result of a certain set of circumstances, fell in love with this dangerous thing, and then came to the conclusion that this is their life. They really could no longer exchange their aerial work for any other specialty on earth. "The first thing, the first thing is airplanes ..." - these words became their motto in life and work.

Pilots are not be born, as well with sailors. It is important when their chosen specialty "almost by accident" coincides later with the tastes and character of a person, and he becomes obsessed with his work. Here is a process that happened with Proskurov. He was a pilot with all the necessary traits: endurance, courage and good health.

It seems that in May a large group of volunteers returned to their homeland. They were replaced by new comrades. Khovansky replaced Ivan Iosifovich, and the names of others could not be remembered. Proskurov and Ostryakov stopped by at Cartagena to say goodbye. It was a hot day, and the southern sun thoroughly baked. I asked them to stay for lunch with the sailors, but for now I suggested walking along the promenade and chatting. The departing Soviet comrades were just as reminiscent of their homeland as those who had arrived from the Soviet Union. Always in such cases it became a little sad and at least wanted to send a letter "with an opportunity". We stopped at the monument to Christopher Columbus, recalling both the former glory of Spain of the 16th century and its decline. For more than six months we worked together, the pilots were already become familiar to the republican fleet. They gained experience and understood what the interaction of aviation with the fleet means, but other considerations required their replacement.

We walked past the Mall Curro, where the cruiser "Libertad" and the battleship "Jaime I" stood. "Oh, this is my Haim," Proskurov remembered Smushkevich's expression, and the conversation turned to naval topics. "If it were my power, I would definitely appoint both of you to serve in the naval aviation at home," I said frankly. Both of them agreed with me. Subsequently, N. A. Ostryakov actually served in the Black Sea and Pacific Fleets, and I. I. Proskurov was destined for a different fate. After a short service in aviation, he against his wishes, was assigned to responsible work in the apparatus of the People's Commissariat of Defense. “I convincingly asked the authorities not to entrust me with a business that I do not know and do not like but it did not help,” Ivan Iosifovich said bitterly. He, of course, as an honest communist, tried to do everything as best as possible, but the business did not go well and he was again transferred to the Air Force. When the war began, Proskurov was "in disgrace." From the People's Commissariat of Defense he was transferred to Petrozavodsk to command a small air unit. However, the repression did not end there either. For failure in the early days of the war, honest people often suffered. So it was with Proskurov.

I think pilots will write more often about him. He deserved it.

I happened to closely meet N. A. Ostryakov both in the pre-war period and in the first months of the war. He combined exceptional courage with amazing modesty. He enjoyed great authority, although he could not severely punish or strongly scold his subordinate even when it was needed. Tall, with large blue eyes and slightly curly hair, he made a very pleasant impression. I first encountered him when he was a pilot in the detachment of I.I. Proskurov and flew bombers under instructions from the Cartagena naval command. In heroes, courage often coexists with boldness or bravado in everyday life both in and outside of service. Ostryakov was always extremely modest and never talked about himself. In Spain, I happened to observe when the planes of this unit often went at night to the limit of their fuel resources to bomb Granada, Malaga or look for rebel ships at sea.

I met Ostryakov again after returning from Spain, as we had to work together for several years. When I returned to the Soviet Union, Ostryakov commanded a brigade of bombers in Sarabuz on the Black Sea. He was loved there, and the brigade soon became a model compound. Ostryakov led by personal example, flying day and night in the most difficult conditions, which at that time was not easy. Like in Spain, he did not know how to scold his subordinates. His second-in-command, pilot M. I. Samokhin, much later, Recalling Nikolai Alekseevich's work on the Black sea, told me how he sometimes invited him to his house and asked him to call such and such a commander and scold him properly for the violation of flight rules.

One episode was well remembered. After returning to his homeland he engaged in complex combat training at night, Ostryakov once equipped with live bombs, flew out in search of and to bomb target ships. As soon as he got off the ground and crossed the road on a southern dark night, he hooked the wheels on the haystack wagon that a horse was pulling. The frightened driver drove off and did not come to his senses for a long time, and Ostryakov, having damaged the wheels, was forced to abandon flying out to sea and made circles over the airfield. On the ground, they discussed what to do: leave the plane and jump out of it or try to land, having previously freed itself of bombs. The offer to jump out by parachute was rejected by Ostryakov. He went out to sea and, dropping the dangerous cargo, returned to the airfield. The night stood out exceptionally dark. "I will make a landing by turning off the motors at the last moment," said Ostryakov, taking aim at the runway.

"We watched his flight in tension," said M. I. Samokhin. - A spotlight beam illuminated the bent landing gear of the aircraft and, it seemed, landing was impossible. But, having made a couple of test approaches, Ostryakov said: "I'm going to land." Searchlights illuminated the runway, and the plane, dropping, turned off its engines. After that, there was nothing left to do, as, having reduced speed, to sit on the fuselage. With amazing calm Ostryakova saved the crew and the aircraft. We didn't even report this incident as an accident, but attributed it to a simple breakdown. "

Such an episode from Ostryakov's life also speaks of high moral qualities. His deputy was offered service in the Far East. Knowing the difficult marital status of his subordinate, Ostryakov himself went to Vladivostok, leaving in his place the deputy he had prepared. We met again. I then commanded the Pacific Fleet, and he took command of one heavy aviation brigade. "He should command the angels," they joked when something happened in the unit and Ostryakova was accused of insufficient discipline to violators of order and instructions. Perhaps this was indeed Ostryakov's shortcoming, but he was like that.

In mid-October 1941, I saw him for the last time. Ostryakov was appointed to command the Black Sea Fleet Air Force, and he went to Sevastopol. In difficult conditions of the blockaded city, Nikolai Alekseevich led his units into battle. It was said that excessive and unnecessary neglect of danger led Ostryakov to premature death. He was at one of the airfields near Sevastopol, when the bombing began, during which he died, not wanting to go into cover. Together with him, the same wonderful pilot Korobkov died. Being the Deputy Commander of the Navy Air Force, he, like Ostryakov, was extremely brave and loved to joke with death. You cannot win without such people, and their death is especially acute.

I singled out Proskurov and Ostryakov not at all because the other pilots of this detachment are less worthy. Many good words can be said about the navigator Prokofiev, whom I met less after Spain, and about others.

During my stay in Spain, the main aviation adviser was J. V. Smushkevich. We established good friendly relations with him, which continued in the Soviet Union until the tragic death of this brave man.

I had to meet with Smushkevich in Cartagena, when air cargo arrived, and then when interaction of the republican fleet with aviation was required. Sometimes the situation at sea required the intervention of larger aviation forces than those at the disposal of the fleet. First of all, this applied to fighters, who often had to strengthen the defense of Cartagena and neighboring airfields where transport planes arrived. It was then that we turned to Republican Air Force Commander Ignacio Cisneros and General Douglas. We turned to them in cases where the situation on the Levant coast was complicated and the rebel fleet threatened communications and coastal cities; the only real way to strike at the rebel ships stationed in Palma (Mallorca island) was aircraft. It happened that the situation at sea required immediate assistance from aviation and then I had to urgently contact General Douglas and develop a general plan of action. So, for example, an unexpectedly critical situation suddenly appeared in the Almeria region when, in April 1937, the battleship "Jaime I" ran aground and the cruisers of the rebels "Canarias" and "Baleares" roamed nearby.

Smushkevich arrived in Spain in October - November 1936. He stayed in Valencia to leave for Madrid a few days later. After a car accident, I stayed for several days in Valencia, where Yakov Vladimirovich found me. The first shipments of aircraft were already unloaded and assembled at airfields; these were mainly I-15 fighters, which after making several test flights went to Madrid. There was heavy fighting in November. Now a batch of bombers was expected which General Douglas was very interested in.

"We urgently need airplanes he managed to blurt out," said Smushkevich, barely opening the door to my room at the Metropol Hotel. Before that, I only talked to him on the phone once. He turned out to be a stocky, medium-sized man, with black curly hair in a beret and a leather jacket. Joking, he asked how my foot was and if I had long I to stay in a horizontal position. "No, two or three more days, and even on crutches I'll leave for Cartagena," I replied. The atmosphere was really unsuitable for bed rest. There were several transports on their way and then they called and reported that the rebels had allegedly landed troops somewhere in the vicinity of Torrevieja, that is, not far from Cartagena. There were very troubling days associated with Franco's advance on Madrid, with the move of the government to Valencia and with the activity of the rebel fleet. Smushkevich sat down beside the bed and we went on to talk about arriving airplanes. Smushkevich asked to unload and deliver them to airfields as soon as possible, he thought about using these aircraft as if they were already in Cartagena. In fact, the transport "Santo Augustin" was still in the eastern part of the Mediterranean Sea and the most crucial part of the route from Bizerta to Cartagena he had yet to go, and the fleet had to meet him at sea. I realized that General Douglas will be the senior aviation adviser and that it actually depends on him to allocate the right amount of aircraft to help the fleet. To my playful question, how many planes - fighters and bombers - will be given to the fleet for successfully conducting the "game", Smushkevich also answered with a joke. Everything that was expected was already on his account and far from necessary to cover his needs. However, I managed to convince him that the reliability of security depends on the receipt of aircraft, not only arriving on this transport, but also in the future. The Spanish liners “Santo Tomas”, “Magallanes”, “Mar Cantabrico” and others were already loaded in the ports of the Black Sea. In just these days Cartagena was heavily bombed. The airfields where the planes were going were also harassed. At sea, traffic has become more dangerous due to the appearance of the cruiser “Canarias” in the Mediterranean. The security of incoming transports required the Republican fleet to go far towards the shores of Africa. They agreed that Smushkevich would allocate several fighters from the very first batch of planes to cover Cartagena and airfields and I agreed to be patient regarding bombers: they were needed very much on the Central Front. On this we parted. I soon went to Cartagena and Smushkevich to the land front. For several months we did not see him. After the defeat of the Italian corps near Guadalajara where Smushkevich took a personal and most active part, he moved his headquarters to Murcia. It was very close to Cartagena. From here, it was more convenient for him to manage operations in all directions and monitor the organization of new air units near Cartagena.

The battle for Madrid in November 1936, then a short respite in December, the Republican offensive on Jarama and finally, the rebels attempted to bypass the northern direction of Madrid with the help of the fresh Italian corps in the first half of March 1937, all of which are memorable events for Smushkevich.

At the very beginning of March the preparation of a new attack by the Francoists across the Valencian Highway became known by Madrid, according to Douglas. Together with the republican aviation commander Ignacio Hidalgo de Cisneros they concentrated at the nearest airfields to the front an air group of 45 fighters, 15 attack aircraft and 11 bombers. The first rebel strikes were over the ground units and especially the international brigade. The enemy's calculations for an easy victory failed. Although the Italian corps, as we now know with certainty had 50,000 troops, 222 artillery pieces, 148 tanks, 60 aircraft and at least 1,500 different machine guns, it could not capture Alcalá de Henares in order to cut off an important road to Valencia and then start an attack on Haram again.

By that time commanders of republican formations had gained experience and learned to interact with tanks and aircraft. This, perhaps should primarily explain the success of actions at Guadalajara.

Y. V. Smushkevich told how he and Cisneros, on the eve of the decisive offensive of the Italians, coordinated the actions of aviation on March 12 and despite the bad weather were ready for battle.

It was decided not to use the aircraft immediately so as not to reveal their cards to the enemy.

On March 18, the Republican counter-offensive began. It was then that skillful interaction of ground troops with aviation was required; the aircraft commanded by Douglas took off several times a day and smashed the retreating parts of the enemy. The Italians lost 1,500 people killed, 1,200 prisoners, 65 cannons, 13 howitzers, about 500 heavy machine guns. Republicans captured more than 5 million rounds and about 30,000 artillery shells.

I did not always succeed in monitoring the actions of the Republicans on land, but I knew the actions of the aviation quite well. Soviet volunteer pilots constantly came to Cartagena, and all the Chatos, Moskas and Katyushas gathered at the Los Alcazares airport next to us. Our average SB bombers were called "Katyushas" long before the Soviet "Katyushas" rockets began to smash the Germans near Moscow and on other fronts.

Here I cannot retell everything that I heard about Smushkevich. People who fought with him will write about this. I want to tell only about individual meetings with him in Valencia and in Cartagena. So, in March, Yakov Vladimirovich arrived in Cartagena together with Air Force Commander General I. Cisneros, when a large transport had just unloaded at the port and dozens of aircraft were being assembled in Los Alcazares and San Javier. War in the air took on unprecedented dimensions. The lull that came after the unsuccessful offensive of the rebels on the Central Front was used by both sides to reform the old and organize new units. The Germans and Italians sent bombers and fighters to the rebels, and the Republicans did everything possible to quickly assemble and master the aircraft arriving for them: "Chatos" - snub-nosed fighters "I-15" and "Katyusha" - medium-sized bombers that have already proven themselves in combat conditions. Soviet volunteer pilots showed courage. There was a lot to do in Valencia and I often had to go there. Here I met Smushkevich again on 8 Alboraya Street where our land comrades worked. Together with them, he was involved in planning future operations and issues of assistance needed by Republican aviation. This required my presence. The number of new units and the restoration of old combat readiness depended on the successful delivery by sea from the Soviet Union of aircraft, tanks and cannons. The issue of providing fronts with aviation was especially acute, and therefore, having finished our general conversations, Smushkevich and I found a free room and continued a detailed discussion there of our aviation fleet problems. "First of all, it is necessary to deliver planes to Cartagena without loss," I said. As practice has shown some ships, the "players", will not always be reliably covered both at sea and in base. We need fighters, at least for the duration of operations. Previous allocation was clearly not enough. There were 5 bomber aircraft missing. I recall with pleasure how when discussing such issues, I always found on the part of Smushkevich a full understanding of the role of aviation in the war at sea. So it was now. We clarified the questions about how many fighter will be required for reliable protection of the base during the unloading of large transports in it. We found out how far fighters can go to sea in order to ensure the movement of the squadron at the time of approach to the port, when the probability of raids is especially high. Then we discussed the help that bombers can provide to the fleet in carrying out main tasks and, above all, tasks in communications, if it is threatened by large rebel forces - the "Canarias" and the "Baleares".

"Well, what does your Haim not break these cruisers?" Laughed Yakov Vladimirovich, when he clearly did not want to steal from the fleet the small number of aircraft that we had. Smushkevich agreed to allocate an additional number of fighters to the fleet for the duration of operations and to keep some of the bombers ready for two or three critical days when the fleet is at sea.

It was agreed that the five planes permanently given to the fleet under the command of I. I. Proskurov should be used only for attacking rebel ships and the planes given for a time should be sent to bomb enemy bases and ports with ships there. In fact, this did not work. In fact, five Katyushas flew to bomb ships at sea, bases and even railway junctions, and General Cisneros and Smushkevich were only able to single out land aviation for the bombing of Palma on the island of Mallorca. But this is not their fault, but a common misfortune - there were not enough planes.

To coordinate our plans as soon as possible, we decided to go to the ministry, to I. Cisneros. We were lucky. Hidalgo Cisneros was in place. He, like Smushkevich, really did not want to change his plans, and, giving consent to the allocation of aircraft to the fleet, Cisneros asked to wait with a final decision until he reported to Minister Don Indalecio Prieto.

"When will the next transport arrive?" Ignacio asked quietly. "I don't have the exact data yet, not soon," I answered evasively. "But should we go together to Los Alcazares and Cartagena," General Cisneros suggested. "We could solve a lot of questions on the spot," he turned to Smushkevich. They really had something to talk about there, "on the spot." We decided to go.

At the appointed hour of the next day we left for Cartagena. Cisneros, as the host, suggested making a stop at Cape San Antonio, where tourists usually stayed in peacetime. There, according to Ignacio, you can have a great lunch and enjoy the sea. In addition, he clearly wanted to call on his wife, who at that time worked in Alicante as the head of the kindergarten. To go, so we go ... Cars raced along the shores of the Gulf of Valencia. The famous orange groves stretched along the highway. Spring air already filled with the aromas of flowers and at noon it was even hot. In autumn, whole mountains of fruit usually lay along the side of roads. Yellow, ripe, in those years they did not always clean up on time: there were not enough cars and labor. In the meantime, everything bloomed. Small pueblos are replaced by the cities of Cullera, Gandia, Denia, which are busy in peacetime and quite quiet now, during the war. Highway traffic is also small. In villages and cities there are very narrow streets and God forbid meeting a car or a donkey. Between conversations, we quietly drove up to Cape San Antonio and turned to the sea. A fashionable restaurant, small but comfortable, a private beach and even special holiday houses. This place, according to Ignacio, was visited by all tourists passing on this road. It's quiet here now. However, the restaurant acted oddly enough, as if the war had little to do with it. I remembered Santander ... In the restaurant there were a lot of smartly dressed people, there were dozens of luxury cars nearby. “What kind of people are these?” Asked Smushkevich, turning to Cisneros, wondering after Madrid at well-trained footmen and exquisite dishes. “Not everyone is at war and the burdens of war are not the same with us,” he answered. - In addition, there are many foreigners in Spain who hastily left Madrid when the Caballero government moved from there; now they have "occupied" the Levant and have fun here. " I don’t know how many foreigners were there, but only Spanish was heard from all sides, and it seemed to me that the Air Force commander was a little ashamed of what he saw.

In May or June I had to get here again. Almost nothing has changed. There were a lot of people on the beach, although the restaurant became somewhat worse regarding food but there was still a large selection of food and drinks.

We continued on our way. From Cape San Antonio to Alicante, the road is more mountainous with sharp turns; it resembles our Crimean highway. It is located closer to the sea and the mountain landscape is often replaced by the sea. Here is Alicante. Cisneros appeared at home unexpectedly and, all the more, they did not wait for us there - two "blond Companeros". We did not want to interfere with the meeting of Ignacio with his wife, but he did not let us go. Cisneros's wife, Constance de la Mora, came from a noble family, as did Ignacio himself. As a very young girl saw her unsuccessfully married. Soon she divorced, after which she chose an unusual path for her environment, abandoned empty social life and engaged in useful work. Shortly before the rebellion, fate brought her to Ignacio. Contrary to the will of his parents, Ignacio went to aviation, where the aristocracy did not particularly strive. After the start of the rebellion, he remained faithful to the republican government and did not change his views until the end of the struggle, emigrating to Mexico at the very last moment. When these lines are written, Constance de la Mora is no longer alive; it seems that Ignacio came to the Soviet Union twice, remaining our friend.

After drinking a cup of coffee, Yasha and I left, leaving the couple alone. They said goodbye to Constance near the car. Ignacio promised his wife to call on the way back. We moved on and after one and a half to two hours on the seaside road were at the airport of Los Alcazares.

Already from a distance we saw huge white boxes with parts of aircraft. A large airfield was filled with aircraft at varying degrees of readiness. At one end of the airfield there were fighters and bombers ready for a test flight and on the other only boxes with fuselages or motors were just assembled. Work was in full swing. Soviet engineer Alekseev has long mastered the situation in a foreign country and was his man, both here and in Cartagena. It was he who met us, turning first of all to Smushkevich, as his direct boss, with a semi-official progress report. As if guessing the desire of the authorities, he began his message with the number of aircraft already prepared for departure, and then proceeded to the assembly deadlines for the rest. The the air were calm the last few days in and it was possible to work almost round the clock - day and night. Another five to six days, and then the "gray wolf" will not be scary at all. The airfield is temporarily empty.

Y. V. Smushkevich and I. Cisneros were in their native element, they were happy to look around every finished machine. The aircraft did a good job near Guadalajara, it was rightly considered the main "culprit" of the defeat of the Italian corps and had high hopes for future operations. Without interfering in this "alien business", I watched from the outside how much respect the pilots Y. V. Smushkevich and I. Cisneros enjoyed. This was a clear example of the friendly joint work of Soviet volunteers with Spanish friends under the leadership of a well-established commander of the Spanish Air Force and our senior adviser. In a small house on the edge of the airfield, on the seashore, senior and junior aviation commanders gathered. There were no maps laid out, and technical issues, not operational ones, were the subject of discussion. How many planes, where it is desirable to send them as soon as possible, how many pilots are available, how to prepare the Spanish comrades, so that they quickly master our "Chatos" and "Katyushas". Most of the people sitting here I knew. Indeed, almost all of our volunteers passed through Cartagena and this airdrome. Many real surnames were established later, already at home, for example, P. Rychagov, S. Tarkhov, A. Serov, A. Senatorov and others. Meeting some of them several years later in Vladivostok or in Moscow, we were pleased to recall the difficult but memorable times of the Spanish Civil War.

Let us return, however, to Smushkevich. Before traveling to Spain, he commanded an air brigade in the Belorussian district and enjoyed great authority because he flew very well and could always show how to work hard. In Spain, he was respected and loved. Knowledge of their work and courage in combat conditions could not but win the heart of any subordinate. Real authority is not created by assigned titles or appointment to a high position. Only in practice can it be proved that the title was given correctly and that the authorities were not mistaken in the appointment. Everything falls into place especially quickly in combat conditions. So it was with Yakov Vladimirovich Smushkevich.

“Well, let's go to Cartagena,” - getting up, Smushkevich turned to me when all the issues were resolved. He was on the way, since it was impossible to get to Murcia from here except through a naval base. Cisneros, I remember, decided to stay at the airport.

In the old big house of Capitania we were met by the commander of the base Ruiz and Juan Garcia. While making coffee and for its lovers (from Russian) tea, the conversation turned to the events of recent weeks. After the victory at Guadalajara, the unpleasant impression of the February rebel offensive in the south smoothed out. It seemed that the Republicans had enough money and manpower to defeat the enemy, but the enemy was also gaining strength, the intervention of the Germans and Italians was widening.

The conversation perked up when fleet commander M. Buiza and V. Ramirez, a frenzied Andalusian and commander of a destroyer flotilla, entered the room. “Mucha gente!” (How many people are here), - noisy Don Vicente exclaimed loudly while the humble Buiza calmly walked over and taking off his uniform cap began to politely greet everyone in turn.

Antonio Ruiz was a hospitable host and when his guests gathered he carelessly put aside all his affairs and began to entertain friends. He certainly did not regret that the gathering extended towards evening and that airplanes could appear with the darkness. True, this time everything worked out well.

When they sat down at the table, mutual courtesies began. The courtesy of the Spaniards could not be denied. The officer who necessarily scolded his subordinate ended the conversation with the traditional: “Salud”. Even he checked the pass of the sentry, who shouted: "Parar, fuego" (stand or open fire), - he threateningly pointed his weapon, after a minute, not yet convinced that there was a pass, he already said affably: "Salud", and waved his hand. So it was in the first months of the Spanish Civil War. And here the aviation and naval commanders came together. Sailors have already appreciated the role of aviation in modern warfare and have great respect for pilots. They understood that one could be calm on ships if fighters were on duty at a nearby airport. The pilots also understood that without a fleet they would not have received airplanes, and paid tribute to the Fleet Commander and the destroyer flotilla commander present here.

"Companero Douglas should visit the ships," the sailors said, and Smushkevich, in turn, called them not only to visit the airport, but also to "fly on airplanes" for closer interaction.

Suddenly quickly, in the south, it got dark, and the pilots began to rush home. Smushkevich kept his promise and a few weeks later visited the cruiser "Libertad".

Relations between sailors and volunteer pilots continued throughout the Spanish Civil War. The fleet could no longer imagine its combat activities without aircraft. We spoke more than once about this role of aviation with P.V. Rychagov when he was my neighbor in the Far East, commanding the First Army Air Force. I often talked with Ya. V. Smushkevich, I. I. Proskurov and many others on this subject, already working in Moscow.

During leisure hours

War was war, but occasionally it was possible to rest. When the events, which at first stunned everyone, are dragged out, people usually adapt to the situation and even in the most intense months of hostilities find free time and find leisure hours. War with its tensions and dangers day and night is sometimes replaced by calmer days. In daytime it's brighter on the streets, cafes and restaurants are briskly trading, people who have left to the mountains due to bombing are returning to their homes. If several nights pass quietly without worries and raids, people begin to linger in the streets until dark, and then get used to the peaceful routine, as if forgetting about the war. Something similar happens on ships. Sailors and officers who rarely have leave and only during the day receive permission to linger on shore in the evening and less tense routines are set on ships. Only anti-aircraft gunners at any time of the day or night remain in full combat readiness, and even at times they forget about the danger, until the enemy with his sudden raid reminds of himself.

The inherent desire of any person to relax on a free evening was not alien to Soviet volunteer sailors. There was a need to organize their free time. Those who managed to get to know the Spanish sailors closer went ashore with them. However, most of all I wanted to meet with compatriots, chat, and remember my native fleet. Such a need increased during the days of the great Soviet holidays. Then I wanted to be together not only with Soviet volunteers, but also with the whole Motherland. I remember with what pleasure in such cases we gathered around the radio and listened to Moscow. We liked listening to Moscow radio broadcasts and the ringing of Kremlin chimes.

Perhaps, on one of these holidays the idea arose to create a club for Soviet volunteers in which one could spend the evening, listen to music, talk about that and that. S. S. Ramishvili, our adviser to the base commander, eagerly took up this and two weeks later adapted one of the apartments not far from Capitania as a club. Conchita and Consuelo - young Spanish girls - cleaned this room, set up tables, even created some comfort. A visit to the club was directly dependent on the weather. Moonlit nights have always been fraught with the danger of raids, the city was on guard, and personnel remained on ships. But when the nights were dark, and if it was still raining, the city came to life and our "Officer's House" was filled with Soviet comrades. Conchita took out the gramophone and everyone listened with pleasure to the Soviet records. They were sent at my request and to the taste of some comrade in Moscow. Usually these were songs from movies familiar to and loved by the Soviet audience. Among them, special success was enjoyed by those who sang of the sea and were close to the heart of military sailors. "The country sent us far away to the stormy sea" or "Wide is my native country," someone would sing and others immediately joined in. Some Spanish songs or just tunes were also popular. Once in a club a series of records with songs based on the "Korobeinikov"[47] motif appeared from somewhere. There were records in which the beautiful Russian Katyusha was chanted. Just as all Russians know the name of the fiery Carmen, so the Spaniards remember Katyusha. In Spain, there was even a production called "Katyusha Mukher Rusa," that is, "Katyusha is a Russian woman." When our first bombers appeared, they were immediately christened "Katyusha" and, apparently, not in honor of the Katyusha who "went ashore," but rather the one that was previously known in Spain. It makes no sense to guess which Katyushas went to bomb the rebels, but everyone liked the song, it was equally loved by Soviet and Spanish friends.

Of the Spanish songs, everyone fell in love with the then-fashionable "Maria de la O". The song talked about a gypsy who possessed everything but was not cheerful - she lacked love.

When there were a lot of people in the club it became fun, everyone tried to shout down the gramophone and each other. The Spanish girls barely kept up with the modest dishes. Although everyone wanted to try a dish here that you won't find on a ship and drink one or two glasses of a strong drink, this was not the main thing. First of all I wanted to talk and exchange impressions. At such hours, the heavy service on ships was forgotten and people had fun as if they had gathered in Sevastopol or Kronstadt, and not in distant Cartagena.

"Huevos fritos" (fried eggs) or "vino viejo" (old wine), the "old men" ordered, going up to the sideboard and winking at the "ends". "Simon, hold for me a place over there in the corner where no bomb will strike," someone joked. The life of the partyy was cheerful, to match his flotilla commander, Valentin Petrovich Drozd. A Capitán de Fragata in Spanish service, he went in a

[47] A nineteenth-century Russian folk song

leather jacket with a zipper and only a black beret, famously shifted to the back of his head, sometimes replaced by a uniform naval cap.

Companero Drozd deftly used a relatively small supply of Spanish phrases and could not only order “rice pilaf with clams,” or “a cognac copy,” but also speak with Consuelo on any sensitive topic. From his "boss" Ramirez, he learned several salty naval expressions and therefore used only the Spanish language when he wanted to express his dissatisfaction. He was one of those who are completely given to service onboard ship and still have fun in their leisure time without losing his head. V.P. Drozd often went to the city with his boss, the cheerful, but too noisy flotilla commander Don Vicente, but he liked to go to the club alone, as if completely free from business. His closest friend was the Capitán de Fragata S. Ramishvili.

Eternally busy, Ramishvili did not feel as an ordinary visitor in the club not, but was “in action”. He could have been called or informed of any kind of trouble at any moment by phone, that “pistoleros” had opened fire in the city, but he could help to ensure a good dinner, when it was already difficult to get groceries. Ramishvili and Drozd used to occupy one table and "arroz de santa pola" (rice pilaf with clams) was their favorite dish.

Sometimes a common table was arranged, and when V. Alafuzov, V. Gavrilov, N. Basisty, S. Soloukhin, I. Eliseev, V. Bogdenko and other volunteers came to the club, there was no end to jokes and humor. If it was shortly after the arrival of the “game” with parcels for sailors, then Conchita and Consuel, in addition to the usual dishes, served canned food, cut smoked sausage, and “vino viejo” was replaced by “Russian gasoline”, as the Spaniards called our vodka. Strong drinks were not abused. Traditional coffee was sometimes replaced with tea and we wondered how the Spanish chef masterfully prepared it.

There were days when our generals could be met at the club: D. G. Pavlov or Y. V. Smushkevich, but they usually appeared in

Cartagena for a short time in connection with the arrival of large consignments, gave necessary orders and quickly left.

Pilots from neighboring airfields I. I. Proskurov, N. A. Ostryakov, G. M. Prokofiev and others also came to Cartagena. Less often tankers from Archena stopped by. When wounded volunteers gathered in Cartagena in anticipation of "an opportunity", then those who were "afloat", that is, freely moving, also liked to go to the club.

All volunteers lived in the interests of republican Spain. Victory was celebrated after successful operations at sea or on land fronts; bitterness of failure was experienced with the Spanish comrades. For example, every successful arrival of transport was noted. And after the fall of Malaga in February 1937, visitors to the club were clearly in a depressed state. And, on the contrary, after the defeat of the Italian corps near Guadalajara, in March, the mood was elevated by both Soviet comrades and the Spaniards.

"When are you going to the sea to transport us to Sevastopol?" - half-jokingly, the tank crewman half-seriously addressed the sailors, who were already tired of "resting" in Cartagena on the transports "Magallanes" or "Mar Cantabrico", which were standing at the pier, waiting for an improvement of the situation at sea and at the same time repairing engines before the required sailing. "It's better to slowly get to Sevastopol," the sailors answered, "than hurry up and end up in Cadiz." Such a danger at that time really existed. For example, off the coast of Algeria, the rebels sank the Soviet transport "Komsomol". The entire crew, headed by Captain G. A. Mezentsov, fell into the dungeons of Franco. He was carrying civilian cargo and not even to Spain and yet he was captured. Moreover, it was impossible to risk the fate of people traveling on Spanish transport. However, Cartagena is a place that is not very suitable for recreation, and we took all measures to quickly transport the tankers to their homeland.

If there is calm in the air the sailors drag out the evening up to 11-12 hours after which everyone went to their military posts. It is dark outside. Gloomy massifs stand out against the backdrop of the sea of Mount Sineso and Negrate. The eyes gradually become accustomed and begin to distinguish between the borders of the harbor with the dark silhouettes of the ships. Anyone who is not well-versed in the port and the berths is trying to team up with one of the "old men" to get to their destroyer or cruiser. Worse, if the alarm starts, then the nearest refuge in Capitania serves as a shelter. Conchita and Consuela, not having time to remove the dishes from the tables also rush there. The first shots of anti-aircraft guns are deafly heard from the sea from which the rebel planes like to attack the base.

On Sundays and public holidays the club opened in the morning but was rarely visited during the day as all Spanish sailors went to the city and on the main street of Caglia Mayor together with our volunteers settled somewhere at a table. At that time, the civilian population of the city was also gathering there. Tables taken out onto the street filled the entire roadway and entire families spent their free time behind them. That is the Spanish!

Sitting at the table, the Spanish sailors merrily made noise, discussing the senorita passing by: “Que guapa” (what a beauty), “oh morena” (black eye brows) - and this was in the customs of Spanish cities, especially in the south. The senorita herself considered this a well-deserved praise and was even offended if she was not escorted by such exclamations. She took such remarks for granted and replied: “Gracias, Muchas Gracias” (many thanks). They say about sailors in Spain: one sailor is one hero, two sailors are two heroes, and three sailors are necessarily drunk. This is not entirely correct. During the war, Spanish sailors drank and sometimes got quite dronk, but this phenomenon was not of a mass nature. A more conscientious part of the sailors and non-commissioned officers sought cultural entertainment. Many attended a movie or bullfight (bullfight). The hereditary port tramps with manned the fleet under the king preferred all kinds of entertainment establishments. By the way, senior officers visiting “fun” establishments were considered normal.

Sometimes Russian films were played in the city: "Chapaev" and "Battleship Potemkin". All our volunteers watched these films several times and complained only about the practices that existed in Spanish cinemas: smoking, noise, constant movement of the public, and often scandals. But with their charter, they don't come to someone else's monastery.

Occasionally, our volunteers had the opportunity to go on a mission to Murcia or Valencia. These cities were rarely bombed at that time and a couple of nights could be spent quietly in bed.

In Cartagena there was no "Plaza de Toros" where bullfights take place and therefore everyone who managed to get to Murcia or Valencia sought to get to this performance.

Being in Spain and not watching a bullfight would really be a shame. This was repeatedly told to me by Jose Delmas, the base commander's office administrator. Even as a young man before the war he himself was going to become a toreador and only an accident (a broken leg), by the way not connected with the bulls interrupted his career at the Plaza de Toros.

In love with bullfighting, Jose loved to talk about it. He told how bullfights took place in Greece and Rome, how for more than a thousand years the Spaniards have been "sickness" with this entertainment, and that there is not a single decent city in Spain that would not have a round building like a circus, only without a roof, where the bullfight occurs. He looked with surprise and even regret at those who did not see such an unusually exciting spectacle. Delmas took my word at the first opportunity to go with him to Murcia to watch the bullfight there. How could I not go after that and soon the opportunity arose to carve out a free evening.

Then I happened to see bullfights in Valencia, but the first visit to a bullfight in Murcia was very well remembered. And not because the first impressions are remembered - just the performance in Murcia ended in an incident with one Spanish sailor.

The Plaza de Toros was already full of spectators when we took our places in the amphitheater. For some reason, the performance went according to an abbreviated program and instead of the prescribed three matadors and six bulls; there were only two matadors and four bulls. Otherwise, according to the authoritative statement of Jose, everything went "as in Madrid." Initially, a small orchestra performed the national anthem; in Valencia, usually they still played International. Then, according to all the rules, two picadors appeared on protected horses, they were to tease and at the same time tire the bull.

“But are the bulls also learning to fight, like the toreador?” I asked jokingly at Don Jose. He seriously replied that the bulls were specially prepared for such performances. The size of the bull depends on the class of the matador and the size of the arena. You can’t show a poor example out in the big arena in Madrid or Valencia - it will offend the audience, and it will immediately hiss the matador or even simply expel him. In Murcia, the arena was medium in size and the bulls were clearly not of the first class, but the appearance of picadors on horses showed that both the toreador and the bulls were in top form.

The Picadors angered the already stunned bull. He frantically rushed around the arena, losing strength and acquiring a certain rhythm in running, with stops at regular intervals. After picadors, bandealers with banderillas appeared. Their task was to thrust a pointed arrow (banderily) with colorful ribbons into the bull’s neck with a deft movement. By their severity, the banderillas tear the body of the bull and lead the animal to an even greater fury. Having picked the moment and thrusting their own banderillas, the bandealers can escape into special shelters built on the edge of the arena.

All this is just a prologue, although the colorful clothes of all participants attract the attention of an inexperienced viewer. Then here comes the toreador with a cloak on a red lining and a sword. He is left alone with an angry bull and cannot leave the arena until the bull is killed. The famous toreadors in Spain are especially respected. Often, the temperamental Spanish public throws hats, berets, handbags and other material evidence of their love and respect for the toreador entering the arena. Finally, everything subsides, and the toreador, having drawn his sword, goes towards the bull, who, having run several times past the toreador, throws himself on the red cloak as a bait, then, lowering his horns, stands for several seconds, as if aiming at his opponent to make a jump. For a matador, this is a kind of reconnaissance in battle. Since the red cloak irritates the animal more, the bull is directed precisely at this red spot, and the toreador, taking advantage of this, deftly jumps to the side, holding the cloak with his outstretched hand. Having done this maneuver several times, the toreador goes on the attack and, choosing the moment when the bull is standing, bowing its head, strikes him in the neck. This is by far the most crucial moment of the battle. If the torero succeeded in delivering an accurate and powerful blow, the bull falls instantly, as if knocked down, killed on the spot. It is both disgusting and beautiful. The audience in this case screams, welcoming and approving the toreador. But it can happen differently. When the blow is inaccurate, the wounded bull falls to its knees or, even worse, continues to run around the arena. Here the matador has no expectations for mercy: the public is indignant. The battle ends in this case in different ways: either the matador, who received a second sword, wins but with damage to his glory and finishes off a half-dead bull and leaves the arena embarrassed, it also happens that the bull remains the winner, lifting its enemy upon its horns.

To my question, how many matadors are injured or injured, Jose answered that most of them sooner or later experience the power of a bull, but there are few deaths.

In Murcia, it all ended differently. Not a single matador was injured here, but a sailor from the audience was seriously wounded. Excited by the extra glass of wine, a sailor was unhappy with the excessive caution shown by the toreador and ran out into the arena with a newspaper in his hands, wanting to show how a real brave torero should fight. But he was out of luck. Before reaching the middle of the arena, he knocked down by a bull and carried out in serious condition by the arena assistants, not with the help of the same toreador. This ended the performance. To Jose's question, how I liked this sight, I evasively, so as not to offend my guide, answered that I had watched "con mucho gusto" (with great pleasure) this peculiar sight.

"Still, this is a sight to watch," I decided.

The fascination with bullfighting in Spain is so strong that during the war soldiers sometimes made up this huge crowd of people. True, the fees in this case went to war.

At the beginning of December 1936, weapon transports arrived in Cartagena one after another. The rebels responded with intensified bombing. The arrived planes, tanks and guns were needed in the battle for Madrid. The republican government that moved to Valencia was nervous. Late in the evening, Jan Karlovich Berzin called me and asked me to come to him on Monday morning. I decided to leave on Sunday to slowly prepare my questions. "I'm going to a bullfight", I thought and ordered the cook for the "coche grande" (as Ricardo called our "Hispano Suiza") by 9 o'clock in the morning. This was one of the rare daytime rides. Usually we preferred night driving: an hour and a half faster and also less likely to lie in a ditch. My injured leg still reminded me of this.

"The car is ready," Ricardo reported with his usual punctuality. I drew attention to his sleek "uniform" and carefully combed hair, which usually stuck in a violent whirlwind from under his cap on his left temple. He was really handsome and loved to flirt a little. "Look, Ricardo, how the Senoritas look at you," I joked when we stopped in a provincial town and curious residents surrounded the car. This was actually so: the girls looked with interest at the brave sailor, but Ricardo pretended not to pay attention to them.

At the last moment I found out that the base commander Antonio Ruiz was also called up to Valencia, to the Ministry of Marine and invited him to go together. In addition, my constant companion at that time was my translator and assistant Narciso. I still did not dare to enter into responsible negotiations in Spanish without him, moreover, he replaced the sailor guarding me.

When we left, Ramishvili escorted his commander and wished us a good journey.

The sun was already high and despite December was warming. Clear sunny weather is normal here, cloudy, rainy days are rare. They say there are years when residents do not count even ten rainy days. This is perhaps exaggerated, but close to the truth. In any case, I kept in mind the invariably clear blue sky of Cartagena.

Having decided to go along the coastal road to Alicante, we turned into the suburbs of Santa Lucia, bypassing the city, and immediately left for the "carretera nacional" (asphalt highway). We had all Sunday at our disposal, and we decided to make several stops. Don Antonio knew where you could eat well and spend two or three hours a weekend on the way to Valencia. I relied entirely on him.

"We will stop at Cape Ifash. There is a wonderful restaurant, and in peacetime, tourists traveling along this road must visit this picturesque corner,"said Ruiz. The base commander survived a rather difficult November in Cartagena and believed that he deserved rest.

It suited me. The next "player" was not expected soon, the ships made current repairs and there was a temporary lull in the port. I agreed to use this trip as a kind of excursion. The national highway leading from Cartagena to Alicante along the seashore was not very good, but it also had its advantages. The movement here is very insignificant, and the road goes all the time along the sea. There are no major cities to Alicante itself. Small fishing villages near the beaches and a few salt mining enterprises are all that could be found along the way. Residents here live poorly: few fish are caught, and peasants suffer from a lack of water and are often forced to abandon their land and go to neighboring lead and copper mines. After a hundred kilometers, the picture is changing. The highway turns left, and a small bay opens in front of the eyes, in the depths of which a small, beautiful bay is located. It, like new Carthage (Cartagena), was occupied over a very long time, but it turned out to be less convenient and did not acquire much significance. The fortress built on the outskirts of present Alicante shared the same fate. Alicante most of all reminded me of Yalta. The same small, beautiful, green city is located on the shore of the bay. The same open roadstead and a very tiny harbor, which can hardly accommodate two or three vessels. In stormy weather, sailors are uncomfortable here, and one needs to go to the open sea. I was thinking about this when the car drove into the city, where we were going to make the first short stop. Since I first got there in September, the city has changed. Then it lived a peaceful life, there were many foreigners who fled from Madrid, and tourists accidentally caught up in the war, whom my driver then called all fascists indiscriminately. Now in Alicante there were fewer people: the tourist season was over and business people preferred to be closer to Valencia, where there was a government. In addition, the city was bombed several times. Yet compared to Cartagena, Alicante remained a resort where you could still find a free room in any hotel and eat well in a fashionable restaurant with sea views.

We stopped at a cafe and sat down at a table that was laid right on the street. The waiter accepted the order and immediately left. Our attention was drawn to the silhouette of a warship, which was slowly approaching the harbor. We wondered who this could be. By decision of the London Committee on Non-Intervention, Alicante entered the German zone, but the British, while maintaining diplomatic relations with republican Spain, also often appeared in the Levant ports. Upon entering the port, the ship dropped anchor and a deliberately huge English white military flag immediately became visible. It may have been the destroyer "Hunter", which several months later was blown up by a mine at Almeria. But that day we were not in the mood to discuss purely military issues and exchanging assumptions about the purpose of his arrival, took up coffee. While Antonio had a fun conversation with an acquaintance who approached him, Narciso and I decided to take a walk along the wall, recalling how the "Neva" transported goods there and the Soviet transport "Kursk" on which our fighters arrived two months ago. Now there was no one in the harbor, except for small fishing boats and several patrol boats that returned from the patrol. The sea was calm. The city has not been bombed for a long time. The governor explained this by the fact that he detained about a thousand fascists. Franco knows it will not go well for them in case of attacks on the city. The explanation is naive. The main reason, of course, was the lack of military facilities in Alicante and the lack of strength among the rebels.

When Narciso and I returned to the cafe, Ruiz was already alone and bored. And he was not created for loneliness at all.

"Let's go?" "Yes, of course!"

Moved on. There were about 200 kilometers to Valencia. At first the road went in mountainous terrain, then moving away from the coast, then approaching the sea. Then, from Cape San Antonio, she skirted the vast Gulf of Valencia, on the banks of which we saw the famous orange groves. This road was clearly built to please tourists, who are so numerous in Spain in peacetime.

We, having traveled a hundred kilometers; turn right to the "very blue sea". There were no major cities on the way, and only pueblos were an obstacle to a quick ride. "Maybe it was more profitable to lay roads past these villages in order to facilitate traffic?" I asked when we stopped waiting for a truck to drive away with a small donkey and a tall coach twice as high above it. "Our roads were built under Primo de Rivera by the state, and in the cities it was the responsibility of local municipalities, which apparently did not have enough money," Antonio explained. But the fact remains: the new Spanish-style car drove the national "carretera national" at a speed of 100-120 kilometers, and when it approached a village or city, it wagged along narrow, curved streets, which were also always full of people. The drivers swore, the crowd threw threatening remarks after them, and all the time gained from a fast ride was lost. So it was with us. In addition, on

Sunday, in the Levant, far from the war, life went quite peacefully, and everyone, from young to old, walked along their "central" streets, where the automobile road was passing. In one of the cities there was even such a case: Ricardo tried to overtake a slowly moving donkey with a cart, but did not calculate. His rear wheel hooked the cart and dragged forward. At first, the man sitting on the goats shouted with good curse, then the locals who surrounded us joined him, and we were in the center of events. Some were on the side of the peasant and scolded all the drivers who "did not reckon with the people"; others scolded the narrow streets and sluggish "drivers" with carts. Some looked at us with malicious curiosity, and some let go at us obviously unflattering epithets. Antonio himself had to get out and explain something until the excitement subsided. The cart was dragged a little to the side, and we drove on, but in the meantime, dozens of cars in front and behind were honking frantically, demanding to clear the road, Ricardo, deciding to catch up with the lost time, pressed the pedal as soon as we passed the last house. The hilly foothills gradually gave way to a more even Valencian lowland. The road does not turn as much as before, and Ricardo travels at a speed of 80–90 kilometers.

We did not conduct professional conversations. I tried to learn more about Spain. Thus, I killed two birds with one stone: I recognized the country and trained in the language. My questions, in turn, aroused interest in the Soviet Union in my interlocutor, the Spaniard. I asked why Valencia is interesting. It turns out that the famous generals Cid and Jaime fought here. "So, that is whose honor the battleship of the republican fleet "Jaime I" is named - the pride of sailors and the nest of anarchists," I thought. I learned that more than two-thirds of the famous Spanish oranges are harvested in this area and one-third of the rice comes from fields located on the seashore. Some of the canals in this area are certainly created during the Roman and Arab rule. The Greeks, Carthaginians and Romans left monuments in the form of fortresses and ports from which valuables were exported. Of course, I did not learn from Ruiz that this rich region was inhabited by the poorest peasantry in Spain. The land belonged to landowners and petty landowners. The government of the Popular Front carried out agrarian reform, but the main problems nevertheless remained unresolved.

After driving a little over an hour, the car turned sharply to the right, and we saw the sea. It was noon. The sun was high. The east wind from the sea was gentle and warm. This always happens in coastal areas: in summer the heat is tempered by sea coolness, and in winter the cold recedes. I remember it well in our Crimea: in Simferopol it is winter with snow, and in Sevastopol, and even more so in Yalta in February you can walk without a coat.

We drove up to Cape San Antonio. Actually, there are two capes - San Antonio and Nao. Between them, a small bay with a beautiful beach crashes into the shore in a half-arc. Here, on a picturesque hill, the restaurant is located. A half-kilometer spit with an excellent road leads to it, and a green hill like a float is surrounded on all sides by water. Previously, all the rich tourists who traveled from Barcelona through Valencia and Alicante to Madrid or vice versa were brought here. Now it was not particularly crowded. Daytime, and tourists have become much less. However, half a dozen cars stood at the entrance and half of the tables were occupied. "When the season ends, it's boring here," Antonio explained with genuine regret. But the rules that had been established over the years remained, and the waiters, polite to cloying, quickly scurried between tables. The audience was well dressed. Several Spanish pilots in military uniforms stood out. We are located on the veranda. Ruiz ordered dinner, and judging by the long list of dishes, the menu here did not yet have the restriction that already existed in other city restaurants.

I remembered this dinner, during which we witnessed an interesting everyday scene. We have not yet been served, and we examined the restaurant. Suddenly, the audience began to applaud loudly, and many stood up. I was perplexed. Everyone was looking at the entrance, where a festively dressed group of people appeared. "Que pasa?" (What happened?) I asked Antonio.

"This is a wedding," he replied and promised to tell later, but for now offered to watch.

"Novia, Novia!" Came from all sides. In a wedding dress, a black-eyed girl was moving in front in arm with the groom. They walked to their already prepared table, and praise came towards the bride: "Ay, morena, que guapa!" (Ah, black-browed, what a beauty you are!) She took these words for granted and answered the traditional "Muchas Gracias", bowing.

When the wedding procession ended and everyone sat down again, Antonio explained what this meant. "Perhaps they came from far away," he began, "the ceremonial procession is included in the wedding ritual, the purpose of which is to hear as much applause and compliments as possible at the bride's address." He could not say where this procession came from: from neighboring Denia, a nearby village, or from somewhere from the mountains 100-200 kilometers from here. In a word, it was a "campaign for compliments." "The Spaniards do not want to marry, the Spaniards want praise," the Spanish song says, and what we saw is confirmation of this. True, the novia, in addition to praise of course, wanted to get married. After a few minutes, the bride and groom again caught our attention. At some point, everyone again Applauded and demanded that the newlyweds kiss. Yes, they themselves, apparently, expected this, and therefore the groom did not force himself to beg, tenderly embraced his bride and kissed. This was repeated more than once. We did not wait until the wedding ritual ends, and on the way we continued to talk on this topic.

The car drove into Denia. The city is located almost on the shore of the Gulf of Valencia. Close to the sea we saw rice fields and a little further to the horizon stretched orange groves. In December, the Valencia Valley is not as attractive as in the spring, which begins here very early, or in the fall during harvest. In September - October, on the Levant's roads, I saw mountains of oranges which time did not allow to be picked up. In spring, everything is in bloom, and March is perhaps the most beautiful in the area. Now the car was walking along the road among deserted orange groves and flooded rice plantations. After dinner, everyone was silent; I wanted to take a nap.

Only Ricardo scolded from time to time when he could not move forward quickly, but he also was in a good afternoon mood. The closer to Valencia, the more often there were high advertising signs on the sides of the highway. These were tips to drink the best Gonzales cognac and old wine from Malaga or Jerez. Then the outskirts stretched with shacks and narrow driveways, which soon replaced the wide streets and high houses of the city center.

The huge city with half a million people was now especially lively. The embassies with large staff moved here with the government. True, they said that some missions remained in Madrid, sheltering the Nazis from the "fifth column" in their buildings and expecting Franco to win. But the true friends of the Republic were here. Located on the ground floor of the Metropol Hotel, the restaurant was full of people in the evenings. Journalists from all countries came here to relax and learn the latest news. Various politicians who came to Spain gathered here, regardless of their views and beliefs. For some reason, the Frenchman Malraux was remembered. Nervous, with jerky movements and a cigarette in his mouth. I saw Malraux many times in Spain and remembered again when his last name appeared among the members of the government of the fifth republic. Times change!

We arrived quite early. Calling Berzin, I made an appointment the next day. The afternoon was free. What to do? Rest so relax, and I took advantage of the invitation to bullfight.

There is no need to describe the classic representation of bullfight. Much has been said about this and you cannot write anything new. But the overall impression of this bullfight was much better for me than from the first performance.

More order and everything is set on a grand scale. No one tried to jump into the arena, although there were perhaps more screams; with all this, order was maintained. Ministers sat in the government box. Before the start, the republican anthem and the International were performed. Everyone stood and fighter planes flying in the sky covering the bullfight.

We were already returning to the hotel when we saw a huge crowd escorting the carriage in which the hero of the day rode – the toreador with his beloved. Having finished my business and having a rest in Valencia, I returned to Cartagena.

The death of the battleship "Jaime I"

In June 1937, a sad event occurred.

In April 1937, in the Bay of Biscay near Santander, the battleship "Espana" of the rebels exploded on a mine. For the personnel of the republican fleet and especially for officers, this event caused a dual feeling. The death of the enemy ship of course pleased. However, a ship of the Spanish fleet was lost with which many had a memory of serving on it or joint sailing as part of a squadron and therefore its death caused regret. In this regard, I recall how, examining the besieged Oviedo in Asturias in September 1936, I heard a Spaniard say that rebels sometimes take refuge in buildings that the Republicans do not shell, sparing them as historical value; he added: "It is not easy for the Spaniards to destroy national values and beautiful architectural monuments. The rebels reckon less with this."

In May, the rebels, as if in response to the death of their battleship increasingly bombed Almeria, where "Jaime I" stood. The anti-aircraft defense there was rather weak and two bombs hit: one in the battleship, the other between the hull of the ship and the wall. It became obvious that the pilots "Increasing Accuracy" will now try to finish the job. "It seems to me that the battleship should be transferred to Cartagena," Buiza once said. I supported this proposal, knowing that the anti-aircraft defense of this base was significantly enhanced. The battleship was soon transferred and delivered stern to the Curro pier, not far from the cruiser "Libertad". "Jaime I", an old ship, was difficult to use when the squadron was busy providing communications. The strong influence of the anarchists on him brought a lot of concern to the command. We had to wait for better times to use this battleship. But it happened differently. On June 7, 1937, at about noon I heard a strong explosion, at first I thought about the possibility of a sudden raid from the sea. At this time, Cartagena was bombed by airplanes from airfields in Mallorca and the sea approach was more convenient for them. Entering the duty room, I saw a tall column of smoke rising above the "Jaime I" and no firing at air targets. There was no air alert either. Some residents fled to the shelter, others remained in place, stunned by the terrible sight at the roadstead. Soon everything became clear. An internal explosion occurred in one of the magazines of the battleship, and now other explosions should have been expected if decisive measures were not taken. After a few minutes, a series of weaker explosions followed. This was exploding anti-aircraft shells, laid out near the guns on the upper deck. Tugs poured water into the ship using their powerful fire fighting equipment, the battleship gradually tilted to starboard. All the ships in the port remained in their places, and only "Libertad" was ready to change places just in case, fearing a general detonation of the magazines on the battleship. As usual, in such an environment, one council followed another. Some offered to take the battleship into tow to a shallow place; others considered the cruiser "Libertad" with its fleet staff on board in the immediate vicinity of the burning battleship to be dangerous, while others advised that all attention be focused on saving the battleship itself. M. Buiza was in place, and the correct action was not to interfere with his command. There are times that are not very suitable for advice; in such cases, one must either act or not intervene. Deep sadness on all faces without exception could be read in these minutes, especially when the boats came with the first wounded, and the remains of the dead were transported ashore.

When the half-flooded battleship sank to the bottom with a slight roll to starboard and flooded magazines it no longer posed a danger to other ships and everything gradually calmed down. Only at the scene of the disaster rescue teams worked for a long time.

They did not think about the causes of the explosion. One thing was clear: on the battleship that day the ammunition was overloaded and the explosion occurred at the time of break, when the team was having lunch. However, the death of the largest, although not the most combat-capable ship, acquired great significance. About 80 people killed and 150 wounded exacerbated the case. The next morning Prieto flew in, and we went with him first to the hospital, where the wounded were, and then to the pier, so that we could take a closer look at the destroyed battleship. The huge 12-inch towers that remained on the surface reminded one of the former power of this ship. The crew had already evacuated the ship, although separate groups of sailors were working on the upper deck among the ship's gear which had been randomly scattered by explosions.

"Que barbaridad" (What barbarism), repeated Don Indalecio several times. Not so long ago, he shared with me the joyful news of the death of the battleship of the rebel "Espana", and suddenly "Jaime I" sank in his own harbor.

The most likely causes of the death of "Jaime I" were: sabotage by the rebels, who clearly sought to destroy this ship, or a negligent attitude towards the ammunition on part of the crew. Lack of proper discipline and order on this ship does not exclude both. Prieto and Buiza were inclined to the first reason.

News of the death of "Jaime I" appeared in the press of the whole world. There was no time for sadness. A few days later the squadron went to sea to meet another transport.

After some time, Commander Buiza and I passed the flooded battleship along the Curro pier. He looked sadly at towers and masts sticking out above the water. We talked about the fate of ships. Some have a short but brilliant century, others have been afloat for a long time, and they perish as senselessly as "Jaime I".

"Have you ever heard of the Spanish ship Reina Mercedes?" Miguel asked me. "The first time I hear such a name," I replied.

"Constructed back in 1880, simultaneously with two similar cruisers with a displacement of about 5,000 tons," Buiza began, "she was launched in 1887 and had three huge masts for sailing equipment, although steam engines were considered the main propulsion.

In 1898, when the Spanish-American War began, this cruiser was part of the squadron of Admiral Cervera in Santiago de Cuba; in its condition, it could not steam and was used as a block ship to guard the entrance to the harbor. When the war began, it was scuttled to block the entrance of the American Navy. After the war, the Americans refloated it and still use the ship (1937) as a floating barracks for the cadets of the US Naval Academy. It is curious, said Buiza, that this ship was once again able to raise the Spanish flag under unusual circumstances. It happened as follows: 22 years later, in 1920, when Spanish warships visited the United States, they also went to Annapolis. When the battleship "Alfonso XII", later renamed "Jaime I", saluted to the Americans, the return shots were fired from this once-famous Spanish cruiser. Saluting, she, like a curiosity, had to raise the Spanish flag. But it lasted only a few minutes."

Now I understood the story of the commander. Before our eyes, half-flooded, the very "Alfonso XII" lay on the bottom.

In an American magazine I read that the former cruiser "Reina Mercedes" served until 1957. 70 years after being laid down, it was finally sold for scrap and ended its existence. Yes, ships have different fates!

Homecoming

At the Distant Meridian

Call to Moscow

In August 1937, I received an order to leave for Moscow.

The desire to visit my homeland in Moscow was great but nevertheless, I sadly left the republican fleet and especially my work colleagues. Having hastily gathered myself and not even giving the necessary instructions in case I could not return, I left Cartagena on the appointed evening. As always, cruisers were stationed at Muelle Curro, and destroyers were in the arsenal and at the port's berths. The city was darkened and driver Ricardo slowly moved in the dark along the narrow streets of the suburbs of Santa Lucia until we reached the main highway.

I did not have a clear idea of the purpose of my summons. "For personal information," it was indicated in the telegram. The Republican fleet was busy providing communications and was doing its job well these months. Least of all did I expect that I would not return to Spain anymore. Saying goodbye to my comrades and V.A. Alafuzov, who remained for me, I said that in two weeks I would definitely return to Cartagena. Only in Moscow did I find out that there was an unwritten rule - do not keep our advisers in Spain for a long time. Thinking over this later, I came to the conclusion that it is impossible to approach the replacement of rank-and-file pilots or tankers and chief advisers equally. And here's why: an ordinary pilot or tanker, having arrived in Spain, manages to get accustomed in six months or a year, gain enough experience and participating directly in battles needs to be replaced. The responsibilities of senior and especially senior advisers are much more complicated. These people build all their work on relationships with their Spanish comrades. For effective work, they need not only to get acquainted with the situation on the map, but also to get comfortable, to know the Spaniards with whom to work, to gain credibility. After a year, the advisers begin the period of the most fruitful work and according to the established rule, just then the time comes to recall them to their homeland. So it was with me. The first months of working with Spanish sailors, it took me to get to know people, the situation, and to learn the language. It was only after a joint campaign with the commander of the Republican fleet and the commanders of the formations on ships to Cantabrico and several operations in the Mediterranean, i.e. by the spring of 1937, that I felt that I was doing good. My relationship with M. Buiza and I. Prieto by this time had acquired a businesslike and to the extent necessary, friendly character.

Together with my Spanish comrades I gained experience and they could not blame me for not being familiar with the specifics of the struggle in their country.

When a new Soviet comrade arrived, the Spaniards naturally looked at him with some caution. One cannot ignore the proud nature of the Spaniards. They gave me an example: a little earlier before Y. V. Smushkevich was recalled to the Soviet Union. He was recalled just at the moment when he had unquestioned authority. He participated in the battles near Guadalajara and worked perfectly with the commander of the Republican Air Force I. Cisneros. He only would have worked for the mutual good. But he was recalled. Lopatin arrived instead. And I myself had to hear Cisneros express his regret over the departure of Smushkevich and undisguised uneasiness as he works with a new man.

To a certain extent this also applied to me and a little later with Stern and other Soviet comrades.

However, then in August 1937 having left Cartagena I was thinking about something else.

The martial condition of the Republicans was then stable. Malaga's spring defeat was offset by the defeat of the Italian corps near Guadalajara. Although the enemy was standing at the walls of Madrid he had no chance of seizing the capital. Franco launched an attack on the Northern provinces: Basque Country, Santander and Asturias, and the Republicans were preparing for active operations on the Aragon Front. The scales hesitated not giving a decisive advantage in that or the other direction. The Republican army learned to fight, turned into a massive and organized force. Brave and talented military leaders emerged, such as Lister, Modesto, and many others. True, the Republicans did not have the necessary unanimity in the government but war leadership still improved significantly. Without help from Germany and Italy the rebels would not have lasted two or three months. It was no secret to anyone that Germany openly helps Franco by sending him advisers, troops, and weapons. The Germans were good advisers; their pilots and tankers fought well. Mussolini sent division after division to the peninsula. We joked, and then quoted someone that Italians were created so Austrians could triumph over someone, but they could not be discounted.

The non-intervention committee was just a screen for deceiving public opinion.

And one more circumstance made itself felt. Within the country, the process of broad field cultivation was proceeding rapidly. The credibility of the Communist Party increased every day, and its slogans became popular among the people. This scared not only the conservative government in England, but the Republicans and Socialists in Spain itself. Continuing the struggle with the rebels, they began to increasingly look back at their rear. Now we know how events developed further.

However, it was not these general questions that interested me the most. I had to prepare "information" on the activities of the republican fleet.

The meaning of my report was as follows. A squadron of republicans consisting of 3 cruisers, 15 destroyers with attached planes and torpedo boats has so far successfully coped with ensuring the transport of weapons. However, the recent meetings of the "players" have already demanded the release of the fleet in full force. The rebel fleet replenished with destroyers and submarines "bought" from Italy. He relocated by Gibraltar and began to stand more often in Palma on the island of Mallorca. By this time, Italian warships began to show activity. There have not yet been open attacks on republican transports, but the likelihood of this is not excluded.

The conclusion was that a new stage was beginning in the struggle for communications and above all those linking Spain with the Soviet Union. Maritime communications with France passed along the coast and did not cause much concern to me.

I was preparing a report on the help to the fleet that I would like to receive from the Soviet Union. This concerned sending additional torpedo boats, torpedoes, anti-aircraft guns and fuel, the lack of which the fleet already felt.

Of course, a whole section of the upcoming report concerned volunteer sailors. I wanted to talk about our people and their courageous work on Spanish ships, I wanted to ask our marine group to be replenished with experienced surface ship and submariners, in which there was an acute shortage.

So, in thought, quietly we drove through Alicante. Ricardo asked for a short stop, and I used it to eat. Soon we moved on. The mood was excellent. After all, a trip to the homeland is ahead.

Short stop in Valencia. Recent conversations with G. M. Stern and his assessment of the situation in Spain for a report in Moscow. Early in the morning I left for Barcelona. V.A. Antonov-Ovseenko hospitably invited me to spend the evening with him at the Barcelona Consulate and then move on to the French border. The train to Paris departed in the evening. When the lights came on at the border station with Spain an unusual feeling swept over me. For many months, we have become accustomed to darkened cities and even learned to do without light. An evening in Cartagena is darkness and a restless feeling, as if you did not have to hide in a shelter in case of a raid. And suddenly a brightly lit station, passengers peacefully walking on the platform. "The comrades in Cartagena are doing something now," I thought, and looked toward Barcelona. If there were no mountains here the darkened cities of Spain would obviously sharply mark the border. But then a train came up and we were invited to take our places. In those days many people of various nationalities with all kinds of passports traveled across the French border to either side. This of course was not a secret for customs officials and the border guard but with the tacit consent of the authorities no one was detained. If any incident occurred, then a small bribe in dollars quickly exhausted it. My position was more advantageous. I was traveling with a diplomatic passport.

Less than a day's drive and we are in Paris. This time no one met me and I did not need it. I got in a taxi and asked me to be delivered to one of the hotels near Grenelle Street where our embassy was located. "Now you will not find available rooms," the driver said turning to me. An international exhibition has opened in Paris and all hotels are crowded. "Well then, take me to Saint-Germain," I asked, believing that there, from old memory, they would shelter me. I was not mistaken, Carmen quickly found me a room.

I considered it my duty as a naval attaché to immediately appear to our ambassador in France, Y. Z. Surits, whom I already knew a little. He was ambassador to Turkey when our warships visited Istanbul. The ambassador of course could not remember me, then still a young commander, but I remembered him well. However, in the conversation he immediately recalled all the tricks during the stay of Soviet sailors in Turkey in 1928. He listened carefully to me and asked many questions related to the transportation of goods from the Soviet Union. The balance of forces at sea became less profitable for Republicans every day. Maritime communications were unreliable and land borders with France remained the only way to supply the people's army. Y. Z. Surits said that soon Soviet transports should arrive at French ports and measures were being taken to transport goods across the Spanish border. As it is now known, the French government was reluctant to give permission for transit and in 1938 it was completely canceled. Loads were delayed in France while they were urgently needed by the Spaniards. I had to live in Paris for three days. A young man in a Cook office selling tickets to all corners of the world could not provide a seat in a car without a transfer in Berlin before this deadline. Despite my diplomatic passport I didn't want to stop in Germany for a diplomatic passport: Quiepo de Llano had mentioned my last name several times in connection with the Deutschland bombing, and I considered some caution necessary.

When the Nord-Express left the Gare du Nord in Paris, all thoughts turned to Moscow.

I didn't have to go back to Spain. "We need people here," Voroshilov told me, and offered to rest in Sochi for the time being. "We will inform you of the final decision," he added, saying goodbye.

I learned about the subsequent events in Spain from returning comrades or from the Newspapers.

The night battle of the republican fleet with the rebel ships. The death of the cruiser "Baleares"

The relocation of the rebel ships to Mediterranean bases and their more frequent presence on the island of Mallorca in the second half of 1937 forced the republican fleet to try using its destroyers to conduct night searches and attack the enemy ships with torpedoes. This could weaken the rebel fleet and allow the Republicans to more reliably provide their communications, both day and night.

At the end of 1937 and in January - February 1938, up to ten night searches were carried out. However, all attempts to strike at enemy ships were unsuccessful. In two cases, these ships were discovered, but the attacks were unsuccessful. So, on the night of September 17-18, 1937, in the area of Barcelona, the destroyers "Antequera", "Gravina" and "Sanchez" found the cruiser "Canarias" and attacked it, but errors in the use of torpedo weapons saved the cruiser. A practical conclusion was drawn from this: all subsequent searches were accompanied by training;

special sorties by individual ships were made for torpedo shooting and training of commanders in attacks.

By March 1938 the situation was as follows: after the launch of the cruiser "Baleares" the rebels, actively supported by German and Italian ships, blockaded the entire Republican coast from the end of 1937. But a reliable blockade of the coast was possible only in the case of the destruction of the ships of the republican fleet.

Having a significant superiority in forces, the commander of the rebel fleet decided to conduct an operation involving the entire fleet, setting himself the task of destroying the Republican cruisers. 3 cruisers – "Canarias", "Baleares" (flag of the commander) and "Cervera", 4 destroyers bought from Italy – "Ceuta", "Teruel", "Melilla" and "Huesca", and 2 new mine layers – "Jupiter" And "Neptuno" left March 5, 1938 from the base of Palma (Mallorca) and headed for Cape Palos (slightly north of Cartagena). As a secondary task, the squadron was ordered to set up minefields in the area of the Republican base.

On the same day, the Republican squadron consisting of the cruisers "Libertad", "Mendes Nunez" and the destroyers "Sanchez", "Antequerra", "Gravina", "Lepanto", "Lazaga", "Jorge Juan", "Escano", "Uueva" And "Valdes" went in search of the enemy.

On the night of March 5-6, 1938, two squadrons collided, which ended in the death of the rebel flagship, the cruiser "Baleares".

Opponent courses converged, but at o(zero) o'clock. 40 min On March 6, the destroyer "Sanchez", which was leading in its group to the left of the cruiser "Libertad", found three rebel cruisers and, having reported this, fired a two-torpedo volley at them. The first attack by the destroyer "Sanchez" was unsuccessful, and the ships quickly dispersed, having not yet had time to fully assess the situation and establish which enemy they were facing.

At 0 o'clock 50 minutes, when the cruiser "Libertad" received a report from the destroyer "Sanchez" and the enemy cruisers were identified, the commander lay down on the course intersecting with the enemy cruisers taking a favorable position and at the same time approaching his base.

After the first meeting with the republican ships, the commander of the rebel fleet, apparently realizing how dangerous it is to stay between the shore and the republican squadron, changed course trying to keep the republican ships to the north and avoid contact until dawn when the advantages of his large ships will be obvious. At 2 o'clock 15 minutes on March 6, enemy formations approached each other again and the destroyer "Sanchez" heading the group to the left of the cruiser "Libertad", found the enemy cruiser for the second instance and fired its remaining 4 torpedoes at it. Then followed the torpedo salvo of the destroyer "Antequera" (5 torpedoes) and the destroyer "Lepanto" (3 torpedoes).

Almost simultaneously, the cruisers "Baleares" and "Canarias" opened artillery fire on the cruiser "Libertad", which in turn fired on the cruiser "Canarias". Artillery fire in the dark yielded no results and was soon stopped.

Three destroyer torpedoes hit the cruiser "Baleares"; which was sinking in flames. The rebel cruisers, fearing torpedo attacks, hastily retreated, leaving the crew of their command flagship to the mercy of fate and the Republicans, satisfied with what might have been an accidental but brilliant victory, hurried back to their base. There were no attempts to continue the night battle, and the belated sending of aircraft and torpedo boats from Porman yielded nothing.

A group of destroyers - "Jorge Juan", "Escano" and "Valdez" - sent to search the area of the island of Ibiza, was not used, although they had a full supply of torpedoes and tried to find and attack the demoralized enemy. At 7 o'clock in the morning, the republican ships returned to base and the rebel ships returned to

Mallorca on March 6 leaving British ships to save 200 crew from the cruiser "Baleares".[48]

The unusual fate of the destroyer "Jose Luis Diez"

This name of the Spanish destroyer struck in my memory from the first visit to Cartagena at the very beginning of September 1936. I carefully examined everything that was shown in the base and especially the ships. There were two destroyers in the Arsenal at that time, and one of them was "Diez". His gray, and frankly, dirty hull spoke of his long stay at sea and not a single sailor would have reproached for the lack of traditional fleet cleanliness. The same also explained the lack of proper order on the deck. The crew worked energetically, enthusiastically fulfilling the decision of the committee. During the transition of the fleet to the North, "Diez" was no different from other ships and without entering Gijon was sent with its two other sisterships "Lepanto" and "Sanchez" to Santander. Gijon is a small port, and the remaining large ships were barely accommodated in it.

When the squadron received orders in October to return to the Mediterranean Sea, "Diez" remained in Cantabrico. With him the later arrived "Ciscar" faced the difficult task of protecting the coast. Both ships were in good standing and met all expectations. But now we are talking about "Diez." Until October 1937, he went to sea more than once and provided escort for several transports with fuel and food, breaking through the blockade in front of the rebel ships. When the "Ciscar" in the fight for Santander was blown up and scuttled, the "Diez" went to Gijon and only at the very last moment left the port heading for Bordeaux. For almost a year he was engaged in repairs there. The rebels did everything they could, with the help of their supporters in France; to force the commander and personnel to go over to their side or prevent repairs but all was unsuccessful.

At the end of August, taking advantage of the darkness, the destroyer slipped out of harbor heading for the strait. This could

---

[48] 441 out of her crew of 1,206 were saved.

not go unnoticed and the large rebel cruisers "Canarias" and "Cervera" were already on duty at the entrance to Gibraltar. Several gunboats supplemented this sentinel chain and German ships were ordered to report in the event of the discovery of a brave republican destroyer.

On August 27, the destroyer "Jose Diez" boldly tried to break through the strait but in the battle with the cruiser "Canarias" suffered great damage and was forced to go into English Gibraltar. The ordeal with repairs began again. At the cost of great effort, the personnel, with the help of the British managed to bring the ship into a combat-ready state and at the end of December a "Diez" exit was appointed for joining the republican fleet.

Not only the intelligence and reconnaissance agents of the rebels, who were enough in the area, but also a simple observation from Algeciras allowed the enemy to accurately determine the day of the destroyer's departure. Several submarines were previously concentrated in the Bay of Algeciras, literally next to the English Gibraltar.

On the night of December 28-29, 1938, when the situation in Spain was extremely difficult, the destroyer "Diez", true to its duty decided to go to sea, accept battle and try to break into its bases.

Leaving the bay and not having time to develop a full speed, the destroyer entered battle with rebel gunners and, having rammed one of them ("Volcano"), received several hits in the engine room. Having lost control, he ran aground on English territory and again fell into the hands of the "neutral" command of English Gibraltar. After some time, the "Diez's" crew was interned by force and after a long ordeal achieved a return to republican Spain, the British handed the destroyer over to Franco.

The clash of the republican fleet with the rebels in March 1938 and the sinking of the cruiser "Baleares" show that, despite the superiority of the rebel forces at sea, the republican fleet was ready to continue the fight. Republican sailors still carried out the task of protecting the coast, preventing landings. However, the general situation on the fronts and the policies of the Western powers (England and France) did their job. A capitulant conspiracy was brewing among politicians - republicans, socialists and anarchists, as well as among senior officers in the army and navy. The capitulators did everything in order to stop the struggle, to prove its meaninglessness. The landing operation in the Motril region, scheduled in December 1938, which was supposed to ease the situation of Catalonia, was disrupted under the pretext of a lack of troops and fuel. Individual ships, like the destroyer Diez", were still capable of desperate attempts at resistance but general tiredness and defeatist sentiments spread more and more. The fleet now stood more in Cartagena and only a few ships went on risky operations to escort transports. The situation in Cartagena itself was also getting worse. Returning at this time to his post as fleet commander, Buiza at a meeting of the high command convened by Negrin on March 1, 1939, openly sided with the capitulators. Fascist elements intensified and on March 4 a rebellion broke out in Cartagena. This was suppressed but served as a pretext for the withdrawal of the fleet by order of M. Buiza to Bizerte (Tunisia), which ended the activity of the republican fleet. This happened on March 5–7, 1939.[49]

* * *

The main achievement of the republican fleet was the successful security of communications (and above all with the Soviet Union) in 1936–1937, without which it would be impossible to form a new republican army and wage war on all fronts.

There is also no doubt that the fleet, despite the lack of officers, the lack of fuel and ammunition and the disadvantageous geographical position of its only base Cartagena could have done much more, especially in the first period of the struggle, when the rebel fleet was weak. This is due to the shortcomings of the fleet leadership in the war as a whole on the part of the high command

---

49 With some irony White elements of the Russian Black Sea Fleet fled to Bizerte in 1921.

and government. In addition, it must be borne in mind that the assistance provided to the republican government with aircraft, tanks, artillery and other equipment least affected the fleet. The Republican fleet was forced (unlike the rebels) throughout the war years to be content with the naval staff that remained on the government side. Due to the specifics of the fleet, even weapons and equipment on ships (guns, torpedoes, etc.) could not be replenished from the outside. England, whose help many Spanish ships were built, could provide the necessary assistance to the Republic but it did not.

As for military experience, the role of aviation for any operations at sea and the importance of air cover were first revealed in Spain. It became clear how important it is to have aviation in the fleet under a single command and train it in interactions with ships. Frequent rebel air raids on ships in the base and during the campaigns revealed the role of air defense and the special significance of fighter aviation in it.

The experience of the Spanish war was not lost on us. The suddenness of the raids, especially at the beginning of the war, made us think about the need for constant combat readiness of the fleet, and this was taken into account in the early days and hours of Nazi Germany's attack on the Soviet Union.

List of Illustrations

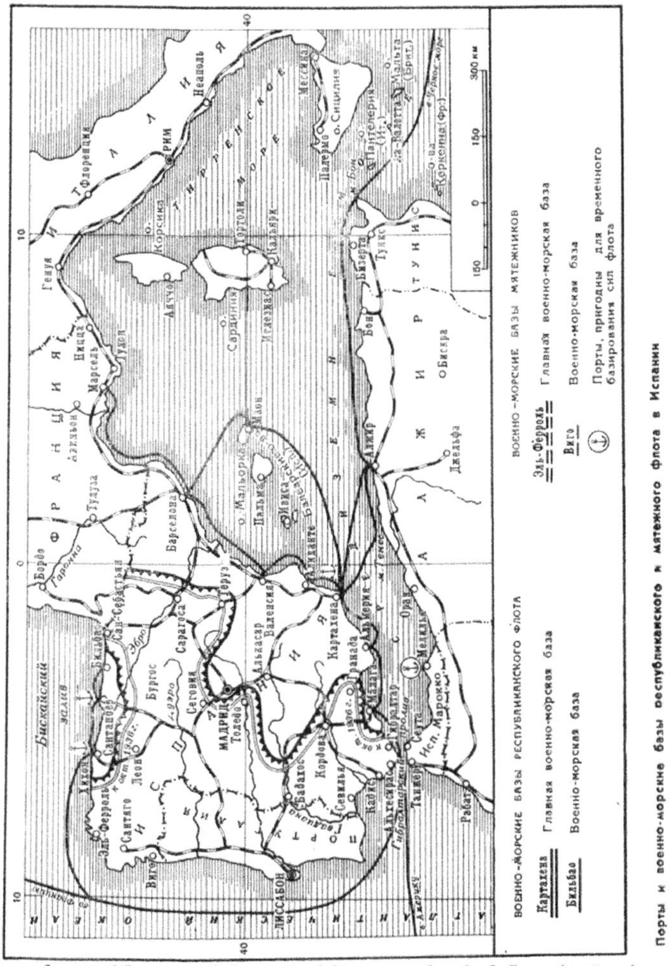

Ports and naval bases of the republican and rebel fleet in Spain

Dolores Ibarruri

Indalecio Prieto

N. G. Kuznetsov, B. F. Sveshnikov, A. Ya. Byaller and A. V. Yumashev in Paris

V. Gorev

Manuel Azana

Cartagena (view from Pier Curro)

Destroyer “Sanchez”

Battleship "Jaime I" after the explosion

"Viva la Republic!"

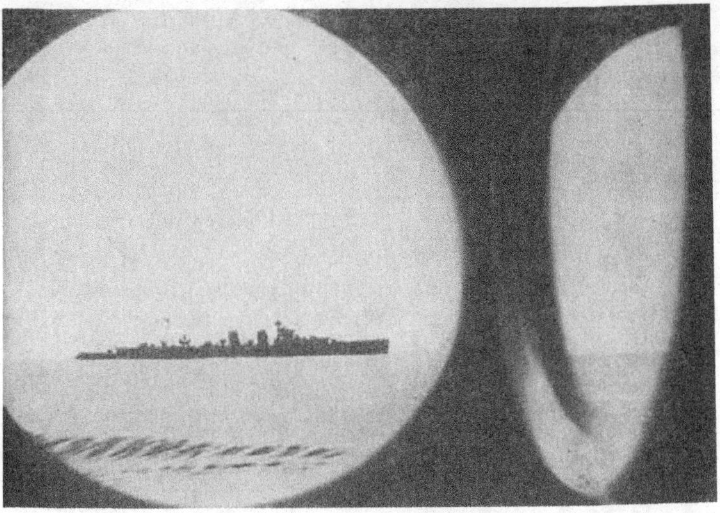

Cruiser "Libertad" on a Campaign

Destroyer "Lepanto"

Battleship “Jaime I”

During leisure hours

The cruiser Mendes Nunez

After shelling the coast

N. G. Kuznetsov and G. M. Stern

S. M. Krivoshein

Juan Negrin

S. S. Ramishvili

VA Alafuzov

V.P. Drozd (left) with a Spanish friend

N. E. Bassist

N. N. Voronov

International Brigade

Soviet tanks near Madrid

N.A. Ostryakov

I. I. Proskurov

The destroyer "Diez" after the battle

NP Egipko

G. M. Prokofiev

Soviet aircraft I-15 (Chatos) in Spain

Y. V. Smushkevich among the fighters

Ignacio Cisneros

S. Tarkhov

Y. K. Berzin

The death of the "Baleares"

Made in the USA
Monee, IL
05 November 2026

67808318R00154